T0122846

The Problem with Pilots

The Problem with Pilots

How Physicians, Engineers, and Airpower Enthusiasts Redefined Flight

Timothy P. Schultz

JOHNS HOPKINS UNIVERSITY PRESS   BALTIMORE

 Published 2018
Printed in the United States of America on acid-free paper
9 8 7 6 5 4 3 2 1

Johns Hopkins University Press
2715 North Charles Street
Baltimore, Maryland 21218-4363
www.press.jhu.edu

Library of Congress Cataloging-in-Publication Data

Names: Schultz, Timothy Paul, 1966–, author.
Title: The problem with pilots : how physicians, engineers, and airpower enthusiasts redefined flight / Timothy P. Schultz.
Description: Baltimore, Maryland : Johns Hopkins University Press, 2018. | Includes bibliographical references and index.
Identifiers: LCCN 2017026551 | ISBN 9781421424798 (hardcover : alk. paper) | ISBN 1421424797 (hardcover : alk. paper) | ISBN 9781421424804 (electronic) | ISBN 1421424800 (electronic)
Subjects: LCSH: Fighter pilots—Effect of technological innovations on—United States. | Aviation medicine—United States. | Aeronautics, Military—Technological innovations—United States—History—20th century. | Aeronautics, Military—United States—History—20th century. | Airplanes, Military—Technological innovations.
Classification: LCC UG625 .S35 2018 | DDC 616.9/80213—dc23
LC record available at https://lccn.loc.gov/2017026551

A catalog record for this book is available from the British Library.

Special discounts are available for bulk purchases of this book.
For more information, please contact Special Sales at 410-516-6936
or specialsales@press.jhu.edu.

Johns Hopkins University Press uses environmentally friendly book materials, including recycled text paper that is composed of at least 30 percent post-consumer waste, whenever possible.

Contents

Preface

> But the fact is the same, that by reason of feebleness and sluggishness, we are unable to attain to the upper surface of the air; for if anyone should come to the top of the air or should get wings and fly up, he could lift his head above it and see, as fishes lift their heads out of the water and see the things in our world, so he would see things in that upper world; and, if his nature were strong enough to bear the sight, he would recognize that that is the real heaven and the real light and the real earth.—Socrates, in Plato's *Phaedo*

In traditional portrayals of combat aviation, the pilot reigned supreme. He emerged from the First World War as the icon of flight, confident master of throttle, stick, and rudder—a heroic figure in a leather jacket and white scarf. Yet this image became a mirage. Dramatic improvements in aircraft performance exposed the human operator as feeble and unreliable. Pilots were unable to function effectively at high altitudes, against excessive G-forces, or in clouds or darkness. They were an engine of fatal errors. The problem with pilots was that they became the weakest link in manned flight.

In the three decades after the Great War, various experts swarmed around the airman–aircraft complex and developed elaborate means to overcome human physiological and cognitive limitations. Among these experts were flight surgeons who specialized in the maladies that piled up as aircraft flew higher, faster, and farther. In collaboration with engineers, these surgeons produced life-support technologies that enabled the human machine to function in environmental extremes. They also discovered that all pilots lacked the ability to remain spatially oriented in inclement weather, and a new paradigm of flight emerged where special instruments replaced the pilot's dangerously inaccurate seat-of-the-pants instincts. These efforts connected the technological, the medical, and the human element and helped transform the airman–aircraft relationship.

Along with physicians, aeronautical engineers gained greater status in military aviation as they developed new means to relegate various pilot functions to superior forms of machine control. The gyro-stabilized autopilot, for example, usurped some of the pilot's traditional duties, and the air force altered the pilot's role as it sought to integrate human and machine components for optimal system performance. Pilots became a biomechanical component in a complex cybernetic system that placed increased reliance on automation. The development of unmanned military aircraft between 1918 and 1945 further challenged the notion of pilot supremacy and relevance, a phenomenon at full throttle today. Such innovations eroded the pilot's position as the central, most essential control element and elevated the role of physicians and engineers.

The fundamental changes in the airman–aircraft relationship instituted before the end of World War Two were spurred by airpower enthusiasts who sought to advance the institutional credibility and power of the air force. In order to mitigate the problem of human limits, they fostered new innovations in life-support equipment, instrument-based flight, automatic control systems, and unmanned aircraft. These innovations emerged from a complex interplay between biology, technology, scientific research, and military necessity that redefined flight by transforming the airman–aircraft relationship. This redefinition is evident in modern airliners and jet fighters, where pilots struggle to keep pace with advances in automatic control and machine intelligence. These advances, particularly in autonomous capabilities, pose new problems for pilots and illuminate a leading edge of our changing relationship with technology.

Acknowledgments

Pilots can't fly without a lot of help from others. And so it is with this book. It owes its existence to the work of many devoted professionals who pointed the way forward and upward. Alex Roland deserves special mention as mentor and role model, and Richard Kohn, Seymour Mauskopf, and Margaret Humphreys also showed me the world through different lenses. I am particularly indebted to Steve Chiabotti for creating the conditions where I could meet such people and think new thoughts. He is joined by the faculty of the School of Advanced Air and Space Studies, a concert of professors and friends who taught me much. In addition to the consummate professionals inhabiting the many archives, museums, and agencies crucial to this research, I also thank my colleagues Tom Hughes, Rich Muller, Mel Deaile, J. P. Clark, Mike Weaver, Mike Pavković, John Maurer, Tim Hoyt, Michael Dennis, Mike Sherlock, John Walrond, Robert Noyd, Jim Durando, John Garofano, Dave Brown, John Jackson, I. B. Holley, and many other faculty members of Duke University, the University of North Carolina at Chapel Hill, the US Air Force's Air University, and the US Naval War College for fueling people like me freely with their wisdom. This diverse cast of scholars and educators is joined especially by Phil Haun, who provided the encouragement and runway for this project to achieve takeoff speed.

This book also exists because of the skill and grit of aviators who push to the extremes. Enabled by innovative engineers and physicians, their efforts reveal much about the past and future of our relationship with technology. I am personally grateful to the men and women of the high-altitude reconnaissance community, including pilots such as Al Marshall, Jeff Klosky, and—crucial to the story here—Kevin Henry and Dave Russell. These aviators who dance with the U-2 Dragon Lady at the top of the atmosphere represent the *ne plus ultra* of manned flight, explorers of human limits who marvel at what few others see.

My earlier career as an aviator affirmed that no pilot flies a perfect sortie, and any failings in this book exist solely due to its all-too-human author. This project does not attempt to represent the official position of the US Air Force, US Navy, or any other institution of my affiliation. It is instead my own effort to query the past about the future. Ultimately, I dedicate this to my wife and the real love that never fails, always hopes, and eternally prevails.

Timeline of Landmark Developments (1903–2017)

1903 The Wright Brothers invent manned flight; the onboard instruments were a wind gauge and a stopwatch

1909 US Army Signal Corps buys the first military aircraft from the Wright Brothers

1910 First air-to-ground radio communication

1911 Frederick W. Taylor publishes *The Principles of Scientific Management*, a study of human efficiency that informs human integration in complex systems

1912 US War Department establishes medical standards for aviators

1913 Orville Wright demonstrates hands-free flying with automatic stabilizer

Elmer and Lawrence Sperry demonstrate automatic control via a gyroscopic-controlled stabilizer, and US Army tests first automatic pilot

1914 Congress creates Aviation Section of the Army Signal Corps

1915 Britain creates Care of the Flier service with physical and psychological standards

Congress establishes the National Advisory Committee for Aeronautics (NACA), the precursor to NASA

1916 Army demonstrates airplane-to-airplane radio communication

Lawrence Sperry files patent for automatically stabilized "aerial torpedo"

1917 Army creates Medical Research Board to study pilot health, oxygen deficiency

1918 Army creates Medical Research Laboratory and School for Flight Surgeons for physicians specializing in aviation medicine

US Army Air Service established

Sperry Gyroscope Company produces a gyroscopically stabilized turn indicator

Army Air Service tests a pilotless aircraft, "Kettering's Bug," a gyroscopically stabilized unmanned bomber

US Navy tests the pilotless Curtiss-Sperry "Flying Bomb"

1920 First flight surgeon graduates from Army Air Service flight school

1922 The School for Flight Surgeons becomes the School of Aviation Medicine; graduates first Navy flight surgeons

1925 Major Henry Arnold, future five-star general of the Army Air Forces, predicts growing mismatch of pilot and aircraft capabilities

1926 Army Air Service becomes Army Air Corps

Flight surgeon Major David Meyers and pilot Major William Ocker create paradigm of instrument-based flight to prevent disorientation in weather

Guggenheim Fund established to help "solve the problem of fog flying"

1927 Charles Lindbergh flies from New York to Paris in the *Spirit of St. Louis*

1928 National Advisory Committee on Aeronautics forms Special Committee on Nomenclature, Subdivision, and Classification of Aircraft Accidents

1929 Air Corps demonstrates air-to-air refueling and long-endurance flight

Aero Medical Association created for civilian and military aviation medicine

Lieutenant James Doolittle conducts first zero-visibility flight and landing

1930 *Journal of Aviation Medicine* is established by the Aero Medical Association

Guggenheim Fund disbands; Air Corps continues fund's work on blind flight

1932 Captain Albert Hegenberger conducts first solo zero-visibility flight and landing

1933 Department of Commerce recognizes Air Corps instrument landing system as industry standard

Wiley Post flies solo around the globe in seven days, heavily enabled by autopilot

1934 Air Corps begins development of pressure suits for high-altitude flight

President Franklin D. Roosevelt assigns US airmail duty to Air Corps, and multiple crashes highlight the necessity of perfecting instrument-based flight

1935 Air Corps mandates all pilots complete a formal course of instrument flight training

Harry G. Armstrong constructs Army's first human centrifuge for G-force testing

Air Corps establishes the Physiological Research Unit in the Experimental Engineering Section of the Materiel Division at Wright Field

Air Corps initiates formal program for development of pressurized aircraft

Air Corps Board reinvigorates unmanned aircraft development to reflect advances in automation and radio control

1936 Major Ira Eaker flies blind (instrument only) from New York to Los Angeles

1937 School of Aviation Medicine debuts official journal, *Flight Surgeon Topics*

First flight of an experimental pressurized aircraft, the XC-35

Captain George Holloman supervises the first fully automated approach and landing in adverse weather conditions

Frank Whittle demonstrates first jet engine in laboratory at Cambridge, England

	Air Corps upgrades Physiological Research Unit to Physiological Research Laboratory
1938	Hap Arnold becomes chief of the Air Corps; remains top airman to 1946
1939	FDR authorizes Air Corps expansion; includes ability to train black pilots
	The first jet aircraft, a Heinkel 178, flies in Germany
1940	Boeing 307 Stratoliner, first pressurized commercial aircraft, starts service
	The National Defense Research Council forms Committee on Aviation Medicine
1941	Committee on Medical Research in the Office of Scientific Research and Development subsumes the Committee on Aviation Medicine
	Radar-based ground-controlled approach technology created by MIT physicist
	Air Corps contracts General Motors to develop ten unmanned aerial torpedoes; reclassifies them as "controllable bomb, power-driven"
	The Army Air Corps becomes the US Army Air Forces (USAAF)
1942	Norbert Wiener, father of cybernetics, equates pilots to servomechanisms
	USAAF approves new prototype development of long-range unmanned bombers
	Maiden flight of the first operational fighter jet, the Messerschmitt 262
	Women's Flying Training Detachment and Women's Auxiliary Ferrying Squadron form; become Women's Airforce Service Pilots (WASPs) in 1943
1943	Radar navigation used to guide bombers at night or in inclement weather
1944	USAAF employs VB-1 Azon radio-guided bomb, an early precision-guided mission
	USAAF and US Navy turn "war weary" B-17s and B-24s into unmanned, remotely guided bombers (Operation Aphrodite)
	USAAF employs pressure breathing for high-altitude reconnaissance missions
	Germany fields gyro-stabilized V-1 cruise missile and V-2 ballistic missile
	USAAF contracts creation of 2,000 pilotless cruise missiles (clone of German V-1)
	Hap Arnold creates Scientific Advisory Group led by Dr. Theodore von Kármán
1945	Hap Arnold authorizes procurement plan for 75,000 cruise missiles for use in late 1945–46; production canceled in August 1945
	USAAF establishes Pilotless Aircraft Branch to advance unmanned aircraft technology
	Psychology Branch in the USAAF Aero Medical Laboratory established to emphasize human factors in equipment design
	USAAF publishes *Toward New Horizons* by Theodore von Kármán

1947 Norbert Wiener develops control and communication theory: cybernetics

Congress establishes the US Air Force

USAF cargo aircraft conducts first autopilot-only transit of the Atlantic

Captain Chuck Yeager breaks the sound barrier in the Bell X-1

1949 USAF Medical Service established; includes first USAF surgeon general

A B-50 Superfortress flies first-ever nonstop sortie around the globe

1950 First combat between jet aircraft takes place in Korea

1953 American Medical Association classifies aviation medicine as an official specialty of preventive medicine

Air-to-air infrared guided missile debuts, expanding air combat beyond gunnery

1955 First flight of the U-2, a high-altitude, single-pilot reconnaissance aircraft

1957 Soviet Union launches first satellite, Sputnik

1958 USAF deploys the Northrup SM-62 Snark, the first jet-powered, long-range pilotless aircraft

National Aeronautics and Space Administration (NASA) established

1960 Norbert Wiener predicts machines will transcend limits of designers and limits of human capacity

U-2 piloted by Francis Gary Powers shot down over Soviet Union

1961 Cosmonaut Yuri Gagarin becomes the first man in space (12 April); Alan Shepard becomes first American in space (5 May)

1963 Aviation medicine renamed as aerospace medicine

1967 Theodore von Kármán publishes *The Wind and Beyond* emphasizing need for science and technology to expand the frontiers of aviation

Precision-guided bombs, the television-guided Walleye, first used in Vietnam

1972 Laser-guided bombs produce significant operational effects in Vietnam

NASA develops fly-by-wire system enabling computerized management of pilot inputs; subsequently adopted in commercial and military aircraft

1978 MIT researchers develop control taxonomy for human-computer systems

First Global Positioning System satellite launched, transforming aerial navigation

1981 Long-range, air-launched cruise missiles deployed by USAF

Debut of the F-117 fighter, a highly automated bombing platform

1989 Debut of the B-2 bomber, heavily reliant on computer-based control

1991 Desert Storm aerial campaign illustrates major advances in computational power, satellite communication, and precision bombing

1993 Department of Defense authorizes combat duty for women, including aircraft

1995 NASA identifies shift in pilot's role to systems manager due to flight automation

1996 USAF debuts MQ-1 Predator drone for reconnaissance in Bosnia-Herzegovina

1997 First flight of the F-22, an advanced-avionics highly computerized fighter

1998 USAF debuts RQ-4 Global Hawk unmanned high-altitude reconnaissance aircraft

2001 USAF arms Predator drones with Hellfire missiles

First flight of MQ-9 Reaper, a heavily armed combat drone

2006 U-2 pilot Lieutenant Colonel Kevin Henry narrowly survives decompression sickness

First flight of the F-35, a networked fighter with advanced computer-enabled situational awareness

2008 Secretary of the Air Force predicts drones will soon be 25 to 50 percent of new aircraft

2009 Air France Flight 447 crashes due to breakdown in human-machine interface

USAF chief of staff declares, "We now must reconsider the relationship of man and woman, machine, and air"

Captain "Sully" Sullenberger lands Airbus on Hudson River, significantly assisted by fly-by-wire flight control technology

2010 USAF publishes *Technology Horizons: A Vision for Air Force Science and Technology, 2010–2030*

USAF creates 18X career field for remotely piloted aircraft (RPA) pilots

2012 Number of aircrew trained to fly manned fighters and bombers eclipsed by those trained to fly remotely piloted aircraft

2013 FAA reports overreliance on automation dulls pilot skill and awareness

USAF funds automatic ground collision avoidance systems (auto-GCAS) in F-16/22/35 to correct for pilot loss of situational awareness

Fully autonomous unmanned carrier-launched airborne surveillance and strike aircraft (UCLASS), the X-47B, lands on an aircraft carrier

Navy helicopter squadron includes manned and unmanned aircraft

Unmanned F-16 demonstrates combat maneuvers and supersonic flight

2014 First flight of "optionally manned" Black Hawk helicopter

Safest year ever for commercial aviation

F-16 pilot saved by auto-GCAS in combat sortie against ISIS in Syria

2015 Air Force Research Laboratory's Human Performance Directorate initiates field study to evaluate pilot trust of autonomous systems

USAF publishes *Autonomous Horizons: System Autonomy in the Air Force—A Path to the Future*

FAA outlines rules governing commercial use of unmanned aircraft

Secretary of Navy declares F-35 will "almost certainly be the last manned strike fighter aircraft the Department of the Navy will ever buy or fly"

2016 USAF begins testing an automatic air collision avoidance system capable of autonomously usurping aircraft control from pilot in dangerous scenarios

Defense Advanced Research Projects Agency tests drone capable of autonomous navigation

2017 USAF and Lockheed Martin demonstrate teaming of manned and unmanned fighter aircraft to perform autonomous strikes against ground targets

Abbreviations

AAF	Army Air Forces
AFCE	automatic flight control equipment
AFHRA	Air Force Historical Research Agency
AWACS	airborne warning and control system
BEA	Bureau d'Enquêtes et d'Analyses pour la sécurité de l'aviation civile
CBARS	carrier-based air refueling system
CFIT	controlled flight into the terrain
CMR	Committee on Medical Research
CSBA	Center for Strategic and Budgetary Assessments
DARPA	Defense Advanced Research Projects Agency
DCS	decompression sickness
ECAM	electronic centralized aircraft monitoring
FAA	Federal Aviation Administration
FD	flight director
GCA	ground-controlled approach
GCAS	ground collision avoidance system
GHQ	general headquarters
GPS	global positioning system
HMD	helmet-mounted display
HUD	head-up display
LRE	launch and recovery element
MAC	multi-aircraft control
MOC	Mission Operations Center
NACA	National Advisory Committee for Aeronautics
NASA	National Aeronautics and Space Administration
NTSB	National Transportation Safety Board
PARS	pilot-activated recovery system
PRU	physiological research unit
RPA	remotely piloted aircraft
SAM	School of Aviation Medicine

UAS	unmanned aerial system
UCLASS	unmanned carrier-launched airborne surveillance and strike aircraft
USAF	US Air Force
USAFA	US Air Force Academy
USSTAF	US Strategic Air Forces in Europe

The Problem with Pilots

Pilot Problems and Machine Promises

The most efficient combat aircraft depend little on their pilot. In modern aerial warfare, software and satellites guide bombs to within inches of their target. Electronic equipment detects enemy aircraft long before the naked eye does. During flight through bad weather, special instruments supersede dangerously unreliable human senses, and automatic control systems enable precision flight during surveillance operations or weapons employment. Some aircraft autonomously skim rugged terrain at high speed during total darkness; the pilot rides along. Other aircraft have no pilot at all. Armed, unmanned reconnaissance aircraft conduct many missions previously flown by aircrews. Pilots, moreover, can impede machine performance: physiological limits constrain aircraft maneuverability and shorten mission duration, heavy and bulky life-support systems tax altitude and payload capabilities, and cognitive capacity struggles to keep up with dynamic flight scenarios. Modern combat aircraft thus require a complex, shifting interaction between man and machine.

This book tells the story of how this functional interaction evolved. It demonstrates that scientific and technological developments from World War One through World War Two redefined flight by transforming the relationship between airmen and aircraft. It also illuminates how this phenomenon is embodied in modern commercial and military aviation. To get an idea of the unique, demanding nature of the pilot–aircraft interface and the inherent problems with pilots, consider a recent high-altitude reconnaissance mission that nearly resulted in the pilot's death.

Altitude kills insidiously and quickly. At around 20,000 feet, pilots start to asphyxiate; decompression sickness sets in above 30,000 feet; useful consciousness is limited to a few seconds at 50,000 feet; and at approximately 63,000 feet, blood boils at 98.6°F—body temperature. Despite these hazards, U-2 pilots routinely operate above 65,000 feet. They sit alone in a pressurized cockpit as the reconnaissance aircraft skims along the upper fringe of the atmosphere. The extreme altitude provides a useful perch for the automatic cameras to observe activity below. In case of a cockpit pressure leak the pilot wears a pressurized, astronaut-like space suit, and he or she breathes pure oxygen to ward off hypoxia and decompression sickness. These countermeasures lower the odds of a physiological crisis. While flying in his U-2 over remote territory in 2006, however, US Air Force lieutenant colonel Kevin Henry experienced failure in the technological cocoon meant to protect him from the dangers of high-altitude flight.[1]

Lieutenant Colonel Henry monitored the instruments and performance of his aircraft as it trolled along on autopilot collecting images from the ground. He was comfortable within his life-support technology. At his disposal were water, Gatorade, food, and chemical stimulants.[2] All went well during the first two and a half hours of the mission. Then he noticed the first symptoms of what would become an acute, debilitating case of decompression sickness (DCS).

Airmen at high altitudes risk the same threat as scuba divers who ascend too quickly: as the pressure around their bodies decreases, dissolved nitrogen may bubble out of solution and inflict serious physiological harm. Nitrogen bubbles tend at first to accumulate in the joints, causing severe pain known as the "bends." Divers know that if they swim toward the surface too fast, they'll get bent. Similarly, a couple of hours after climbing through 60,000 feet, Lieutenant Colonel Henry got bent. First, he began to experience pain in his knees. He confirmed that his cabin pressure was holding at a normal 28,100 feet, and he turned a dial on his space suit to increase the pressure around his body. He hoped this would dissolve the nitrogen bubbles erupting in his knees. The pain decreased, and he thought he was OK. Minutes later, however, he felt pain in his ankles; the earlier knee pain measured about "two" on a scale of one to ten, but this rated near "five." It was not a good sign.

In severe DCS cases, nitrogen bubbles may proliferate throughout the body and appear in organs such as the heart or brain. Over the next several hours, Lieutenant Colonel Henry developed a variety of symptoms indicating severe DCS: headache, hot flashes, fatigue, and visual illusions.[3] The illusions took the form of the aircraft rolling in and out of bank, an unlikely event since the U-2 flies on autopilot to ensure stability during image collection. In an effort to counter his growing fatigue, he ate some chocolate pudding laced with caffeine, a decision that would come back to haunt him later.

During this flight, the U-2 was connected by an electronic tether to mission controllers in a Mission Operations Center (MOC) thousands of miles away in the United States. The MOC controlled U-2 sorties through verbal communication with the pilot and a digital interface with the aircraft's navigation system. The pilot played an intermediate role by pressing a series of buttons that tell the navigation equipment to accept the MOC's directives. As Lieutenant Colonel Henry's DCS crisis deepened, this basic cognitive task eluded him. Even as a U-2 pilot with years of experience, he could no longer interpret the navigational display. He informed the MOC that something was wrong. When asked by mission controllers how bad he was feeling, he replied, "pretty crappy." Things soon got crappier.

For the next hour and a half, the mission controllers attempted to give verbal, step-by-step instructions to the pilot in how to navigate toward "Home Base" in the Persian Gulf region and avoid unauthorized airspace.[4] The pilot could not comprehend his current position, maintain proper compass headings, or couple the navigation computer to the autopilot. These were normally the simplest of tasks, but the nitrogen bubbles had crept into Lieutenant Colonel Henry's brain and scrambled his neurological functions. The mission controllers in the United States summoned a local U-2 instructor pilot, Lieutenant Colonel Dave Russell, to help his ailing colleague.[5] At some point Lieutenant Colonel Henry had inadvertently disengaged the autopilot and nearly lost control of the aircraft. Lieutenant Colonel Russell's main concern was to help his colleague get the autopilot back on. Once engaged, the autopilot would maintain safe control of the aircraft and allow for easier navigational corrections. This proved difficult, though, since most of the cockpit instrumentation had become unreadable gibberish to Lieutenant Colonel Henry. It took twenty minutes for Lieutenant Colonel Russell to coach him how to toggle the switches that allowed the autopilot to take control.

While Lieutenant Colonel Russell worked to reestablish a functional interface between pilot and aircraft, the mission controllers also sought advice from an expert on the human machine, a flight surgeon. Flight surgeons are military physicians specializing in flight physiology and aircrew health. The MOC contacted Lieutenant Colonel Henry's flight surgeon at Home Base and relayed his questions to the pilot: How is the cabin pressure? How is the suit and oxygen flow? How do you feel on a scale from one to ten? Do you have blurred vision? Are you nauseous? Nausea poses a unique problem for U-2 pilots since the clear faceplate on their helmet is locked down so they can breathe pure, pressurized oxygen. The trick is to unlock and open the faceplate before throwing up, no easy task when wearing the bulky gloves of the space suit. When nausea hit Lieutenant Colonel Henry, he couldn't get the faceplate up in time before vomiting a quantity of partially digested chocolate pudding inside his sealed helmet.

The chocolaty mess complicated an increasingly dangerous situation. To avoid choking on vomit, his only choice was to open the faceplate. An open faceplate, however, invited grave risk. The U-2 cockpit was pressurized only to the equivalent of about 28,000 feet above sea level, what mountain climbers call the Death Zone. Humans cannot survive in such thin air without supplemental oxygen or weeks of acclimation. Accordingly, U-2 pilots must always fly with their faceplate sealed closed in order to remain conscious. This meant Lieutenant Colonel Henry was in a race to muck out the gunk that smothered his mouth and nose and then close his faceplate before he slipped into unconsciousness in the thin air. He won the race. He was able to reseal his faceplate and feed pure oxygen back into his lungs quickly enough to avoid blacking out. Better to just smell the vomit than to asphyxiate.

The U-2's long descent toward Home Base began shortly after the vomiting episode. In order to descend, U-2 pilots first lower the landing gear to create drag; Lieutenant Colonel Henry, however, could no longer see the landing gear handle. His decaying neurological state disrupted the left half of his vision and created a blind spot where the gear handle should be. Aided by Lieutenant Colonel Russell reminding him where to find the gear handle, he groped around until his hand recognized and lowered the wheel-shaped device.[6] During the descent things got even worse. Once again the autopilot somehow became disengaged, and he entered clouds that obscured his view of the horizon. Normally, flight in clouds or fog requires pilots to use special cockpit instruments that prevent disorientation by portraying the plane's position relative to the ground. This "flying on instruments" requires special training and intense concentration. Since Lieutenant Colonel Henry was unable to see something as obvious as the gear handle, he had little chance of making sense out of a panel of complicated instruments. Its pilot disoriented, the U-2 entered a spiraling descent and plummeted toward Earth.

The out-of-control airplane with its incapacitated pilot sunk through the clouds and emerged into clear skies thousands of feet above the ground. Now able to see the horizon, Lieutenant Colonel Henry mustered enough concentration to return to level, controlled flight. Alerted to the situation, a friendly nation scrambled two Mirage 2000 fighter planes to intercept the U-2. Lieutenant Colonel Henry later recalled one of the fighters zipping right over his canopy, rocking the U-2 with wake turbulence to get his attention. Lieutenant Colonel Russell directed him to follow the fighters toward Home Base.

The descent to low altitude did not improve Lieutenant Colonel Henry's condition. He remembers slumping to the side of the cockpit, his head leaning against the canopy, unable to sit up straight and almost too weak to control the aircraft. The Mirages led him toward the runway, but he was too disoriented to figure out how to land the aircraft and spent the next forty-five minutes performing a variety of erratic maneuvers over the airbase. The weather was clear, but he was in a deep mental fog. He remembered humming over and over the chorus to the popular country-western song, "Jesus Take the Wheel." The only other voice he could hear was Lieutenant Colonel Russell's satellite transmissions from the United States. Unable to tune in the local radio frequency, Lieutenant Colonel Henry could not hear other U-2 pilots at the airfield who could see him and were trying to talk him down via radio. He kept straying away from the runway; the Mirages kept shepherding him back.[7] Lieutenant Colonel Russell commanded him to "Follow the Mirage! Follow the Mirage!"—a surreal instruction indecipherable to Lieutenant Colonel Henry. At one point he remembered feeling a sense of ground rush, seeing dirt and hangars right in front of him, and hearing cockpit alarms telling him the aircraft was in a stall. He boosted the throttle forward and climbed; photographs show his wingtip coming within five feet of striking the ground, and he narrowly missed a large hangar.[8] The U-2 pilots on the ground radioed him to "Eject! Eject!" He didn't.

Lieutenant Colonel Henry was now *in extremis*, very near death from either physiological collapse or crashing into the ground. During these final moments of flight, Lieutenant Colonel Russell tried to convince him that his options were running out. Unless he could regain proper control of the aircraft, Lieutenant Colonel Russell warned, he would need to "step out" [eject] from the aircraft. Lieutenant Colonel Henry could not reply over the radio since vomit had shorted out his microphone. He feared ejecting since he didn't have the strength to sit up straight in the ejection seat. Ejections are a violent affair, and he imagined that firing the rocket-propelled ejection seat would snap his spine if he was slumped over to one side. Finally, he heeded Lieutenant Colonel Russell's commands and managed to string together—largely out of muscle memory—the proper series of physical actions to perform a normal landing. He lined the U-2 up with the runway, reduced power, and glided in for a safe landing.

The aircraft rolled to a stop on the runway. Two other U-2 pilots raced up a ladder and opened the canopy. Lieutenant Colonel Henry was slumped over, unresponsive. They raised the faceplate on his helmet and vomit flooded out. So much for the glamour of flight. Under the supervision of the U-2 flight surgeon, they lifted him out of the cockpit and lowered him onto the runway. The flight surgeon thought cardiac arrest was imminent, and Lieutenant Colonel Henry has a vague recall of someone saying "Get the paddles ready" as he was loaded into a waiting helicopter.[9] An emergency medical team stabilized his vital signs and flew him at low altitude to a nearby hospital where he was treated for shock and near-fatal decompression sickness. Physicians placed him in a hyperbaric chamber to increase the atmospheric pressure around his body and force the nitrogen bubbles back into solution where they belonged. Normal decompression sickness patients require just one session in the hyperbaric chamber. Lieutenant Colonel Henry needed four. Initial magnetic resonance images indicated significant neurological damage in the right frontal lobe of his brain. No one that day predicted he would fly again, yet subsequent brain scans and tests showed steady improvement. A few months later flight surgeons returned him to flying status.[10]

Lieutenant Colonel Henry's episode provides a window on several key aspects of flight. It demonstrates, for example, the reliance on automatic control and specialized instruments. As military aviation evolved, pilots relied more and more on artificial mechanisms for precision flight. The first autopilots appeared shortly before World War One and provided basic stabilization in flight; modern autopilots can operate aircraft during takeoff, climb, cruise, descent, and landing while the pilot rides along. Some automation systems can even wrest control away from an errant pilot to avoid collision. Over time, the masters of throttle, stick, and rudder became managers of switches, dials, and buttons. Notably, a primary goal during Lieutenant Colonel Henry's ordeal was to toggle the correct switches that would return control to the more reliable autopilot. Since the 1920s pilots also came to depend on special instruments for flight in low-visibility conditions. Flight surgeons established in 1926 that gyroscopically based instruments were far superior to human senses for maintaining spatial orientation when flying in clouds, fog, or darkness. Pilots who ignored their instruments in poor weather became hopelessly disoriented and entered a spiraling descent, just as Lieutenant Colonel Henry did when he flew into clouds and was unable to comprehend the data from his instruments.[11]

The Henry incident also illustrates the role of exogenous control in aviation. The Mission Operations Center directed the U-2's flight path from a ground station on another continent. Lieutenant Colonel Russell, the U-2 pilot in the MOC, became in part the de facto pilot and controlled the aircraft by directing the actions of his airborne counterpart (and thus played a key role in saving Lieutenant Colonel Henry's life). Their situation was closely analogous to other ground stations in the United States that fly unmanned aircraft over foreign countries via a high-tech form of remote control. This practice of transferring authority from the pilot in the aircraft to agents outside of the aircraft has long precedent in military flight. Remote control was used to steer aircraft as early as 1922, for example, and by the end of World War Two special radar systems enabled technicians on the ground to guide pilots through inclement weather for a safe landing.

The most obvious lesson from Lieutenant Colonel Kevin Henry's story involves the relationship between machine performance and human physiological limitations. In 1925 Army Air Service major Henry "Hap" Arnold, future commander of the Army Air Forces in World War Two, sketched several graphs depicting aircraft performance improvements since 1911.[12] By World War One, single-seat pursuit aircraft could climb to just over 20,000 feet. By 1925 they operated up to 25,000 feet, and Major Arnold projected altitudes of 40,000 feet by 1930. He noted that such operations must include protections for the pilot such as "oxygen breathing equipment . . . or air-tight compartments."[13] Consistent with Major Arnold's forecasts, the Army Air Service and later the Army Air Corps experimented with cabin pressurization and pressure suits in the 1920s and 1930s. This included the work of flight surgeons who sought to characterize the effects of altitude extremes by experimenting on airmen in low-pressure chambers. Some experiments were designed to induce and study decompression sickness, while others investigated hypoxia, anticipating the risks Lieutenant Colonel Henry faced when he opened his helmet at high altitude to wipe out the vomit. The physiological hazards that plagued him, therefore, were identified and studied by flight surgeons seven decades before his ordeal. Their seminal scientific insights in how the human machine functioned in flight also established that even the most skilled, experienced pilots could not remain spatially oriented when flying in clouds or fog without special instruments. To avoid spiraling into the ground, pilots learned to trust technology instead of their own instincts.[14]

Although pilots enjoyed an elevated status as symbols of airpower between the world wars, behind the scenes they were treated more and more as organic subcomponents of a larger system. They entered the Army's flight school only after meeting an array of physical requirements, and once in flight school their training and performance were standardized so they could become reliable interpreters of instruments, actuators of controls, and followers of checklists. Indeed, once the pilot sat at the controls, he was a machine within a machine. If a pilot could no longer meet physical performance requirements, he faced removal from flight status by a flight surgeon. Moreover, flight surgeons conducted experiments to characterize human physiology in flight and quantify aircrew capabilities. Flight surgeons were engineers of the body, and they collaborated with aircraft engineers to optimize overall man–machine performance. Lieutenant Colonel Henry's normal tasks were fairly mechanical, and he had specific checklists to guide his actions. When nitrogen bubbles disrupted his physiological and cognitive mechanisms, a flight surgeon assessed his interface with the aircraft and attempted to quantify his "pretty crappy" condition.

Lieutenant Colonel Henry's story reveals much about the modern relationship between airman and aircraft. This book examines how the man–machine relationship evolved as Army aviation progressed from the Aviation Section of the Signal Corps in 1914 to the Army Air Service in 1918, the Army Air Corps in 1926, and, finally the Army Air Forces in World War Two. The term *air force* or *early air force* is used here as a generic reference to these predecessors of the modern US Air Force, which was formally established in 1947. This analysis, moreover, limits its scope to the early air force as it examines changes in the man–machine relationship. Although the US Navy and the commercial sector contributed to improvements in aircraft performance during the interwar years, the early air force was the primary locus of fundamental innovations that transformed the pilot's role and redefined flight.

In its original form, flight was an activity in which the eyes, brain, and finesse of the pilot controlled the aircraft and pushed it to its maximum capabilities. In the First World War, aerial combat correlated with an airman's skill. The machines were comparatively simple and the airman's piloting ability loomed large over the power of the weapon system. But advances in aviation science and technology transformed this relationship. As engineers improved engines and airframes, a gap developed between human and machine performance. Pilots could no longer keep pace with aircraft capabilities and military requirements unless they relied on a variety of technological augmentations, such as flight instruments, autopilots, and complex life-support equipment.[15] The air force shifted onto a trajectory toward greater automation, and pilots ceded authority to life scientists and engineers, becoming in many ways servants to technology rather than masters of it.

A wide range of concepts helps assess how the military necessity to fly higher, farther, faster, and with greater precision resulted in new innovations that redefined flight. This includes Thomas Hughes's concepts of technological systems and reverse salients and Edward Constant's notion of technological change. Thomas Kuhn's model of paradigm change also provides insight into the redefinition of flight and the profound effects of instrument-based flight. Norbert Wiener's theory of cybernetics, moreover, provides a framework to understand the trajectory toward greater automation and how human limitations shaped the human–machine complex. These concepts will be explored and applied in subsequent chapters to aid our understanding of the evolving human–machine relationship.

Aviation medicine played a central role in the transformation of flight, and the first two chapters investigate how flight surgeons fostered a practical nexus between aviation technology and science. They explored human limitations and allied themselves with engineers to maximize the potential of piloted aircraft. Chapter 3 examines how flight surgeons and others developed "blind flight," or flight in conditions of limited visibility such as night or fog. Blind flight required a new paradigm of flight in which pilots surrendered their own seat-of-the-pants instincts to instrument-based aircraft control. Only then could airmen function as reliable components in the system of flight. The status of the pilot as an organic cog in a complex cybernetic system receives further analysis in chapter 4, which assesses how automation usurped pilot functions and how engineers perceived and integrated the human element. Chapter 5 describes the air force's efforts to create unmanned bombers in World War One and Two. It also examines how the goal of institutional autonomy cohered with a redefinition of flight that incorporated pilotless aircraft and cruise missiles.

The final chapters accelerate forward to assess how the problems with pilots identified in the first half of the twentieth century shape modern flight. Chapter 6 illustrates how human limitations are mitigated in modern commercial and military aviation. What do the pilots of today's airliners and high-tech fighters really do nowadays? The evidence suggests that the historical transition of pilots from physical masters to system managers is now in full bloom.

Chapter 7 extends this analysis to how sophisticated forms of autonomous technology, including the efflorescence of unmanned aircraft, redefine flight. We shall see in the conclusion the ultimate fate of Lieutenant Colonel Henry and what the future may hold for people we call "pilots."

Overall, pilots were a major problem in aviation development. They fainted, got the bends, or froze at high altitudes. They blacked out during high-G maneuvers. They spun perfectly good airplanes into the ground after encountering clouds, fog, or darkness. They found innumerable ways to let human errors become fatal errors. The innate deficiencies of pilots hobbled the military and commercial potential of aircraft, and physicians and engineers sought innovative means to address this problem and bridge the gap between human and machine performance. This is the story of their work, and it is also the story of how airpower enthusiasts spurred their efforts in order to exploit the air weapon and expand the institutional power of military aviation.

In historian Thomas Hughes's view, the interwar years were a seminal era in American "technological enthusiasm" when inventors, engineers, industrialists, and visionaries created vast technological systems with far-reaching impact. This included automobile production and electric power generation as well as commercial and military aviation systems, all of which illustrated "the organization of complex systems for making war."[16] Similarly, airpower enthusiasts sought to create a technological system that exploited the aerial domain and provided a new means of national power for an air-faring nation. Airpower enthusiasts, like Hughes's technological enthusiasts, "expected the machine to fulfill their visions."[17]

The result was a redefinition of flight: physicians fused pilots with special life-support equipment; engineers created a new reality that replaced a pilot's senses with an array of needles, dials, numbers, and displays; and military necessity required airmen to substitute their clumsy reflexes with superior forms of automatic control. To develop the full potential of aviation and achieve the visions of airpower enthusiasts required a transformation of the human's role. Sometimes it was better to just remove the pilot.

The developments described in this book challenge the traditional legacy of flight that exalted the role of the heroic pilot.[18] Indeed, the early air force redefined flight to such an extent that General Henry "Hap" Arnold, chief of staff of the Army Air Forces, declared in 1945: "One year ago we were guiding bombs by television, controlled by a man in a plane fifteen miles away. I think the time is coming when we won't have any men in a bomber."[19] One of General Arnold's modern successors in the US Air Force, Chief of Staff General Norton Schwartz, echoed this view. When asked in August 2008 if he wanted to build more unmanned aircraft relative to piloted aircraft, he answered: "I think it is inevitable. Now whether fifty percent of the Air Force is unmanned or sixty or forty, I don't have a good sense of that yet."[20] General Schwartz's statement may have startled those who associate aircraft with pilots, but his view was hardly radical since it affirmed a technological trend that redefined aviation before he was born.

The Pathology of Flight

Shortly before the end of World War One, a commentator in the British medical journal the *Lancet* described a pilot as someone who possessed "resolution, presence of mind, and a sense of humor" and was "alert, cheerful, optimistic, happy-go-lucky, generally a good fellow, and frequently lacking in imagination."[1] This romantic view offered little help in terms of assessing the objective physical or psychological characteristics of competent pilots. The article's prescription for maintaining the vigor of such blithely unimaginative extroverts was equally unscientific: "It appears necessary for the well-being of the average pilot that he should indulge in a really riotous evening at least once or twice a month." Perhaps a cheerful disposition and a hardy constitution were somewhat useful, especially since aggressive improvements in aircraft performance during the Great War and subsequent decades levied a heavy tax on the health and performance of pilots.

The efforts to fly higher, faster, and farther supported military and commercial interests, but improvements in speed and altitude strained human endurance and triggered an array of afflictions. In the years leading up to World War Two, three physiological problems became particularly salient: hypoxia, decompression sickness, and sudden blackout. As the pathological effects of flight weakened pilot performance, they strengthened the role of air force flight surgeons. Flight surgeons established and enforced physical and psychological standards for aviators, and the *Lancet*'s unscientific caricature of pilots in 1918 gave way to a more technical appraisal of the human component. Indeed, flight surgeons became clinical pathologists of flight as they developed technoscientific methods to investigate the limits of human function. Their efforts strengthened the nexus between science and aviation, shaped the evolution of the airman–aircraft complex, and legitimized the role of aviation medicine.

"A New Breed of Diseases"

Flight is hostile to humans. It places stresses on body and mind that generate a variety of afflictions, some fatal. In 1937 the assistant chief of the US Army

Air Corps declared flight "a creator of a new breed of diseases," diseases aggravated by dramatic changes in aircraft capabilities.[2] Since World War One, a brisk pace of improvement in aerodynamics and engine power took pilots to both greater height and speed. Higher altitudes promoted fuel efficiency and allowed aircraft to reach distant objectives, plus they could fly over inclement weather and stay out of the range of many types of antiaircraft defenses. Higher speeds provided a maneuvering advantage for fighters and let them catch up to high-flying bombers. Such advances imparted significant military advantage, yet they produced ills such as hypoxia, decompression sickness, and sudden fainting.[3] The inherent dangers of aviation also plagued the psychological health of airmen. Indeed, a whole pageant of disorders awaited them. These disorders required a scientific evaluation of human limitations and spurred the development of aviation medicine.

Consider one of the most common ailments: hypoxia, or lack of oxygen. Hypoxia researchers in 1941 noted that "the human flame" weakens and may sputter out around 20,000 feet.[4] This threat existed even before sleek, all-metal airframes with supercharged engines replaced their slow, wood-and-wire antecedents. As one American aviator in World War One reported: "There were five of us and we ran into five Fokkers at fifteen thousand feet. . . . We climbed up to twenty thousand five hundred and couldn't get any higher. We were practically stalled and these Fokkers went right over our heads [and] tried to pick off our rear men. [Two of us] turned back and caught one Hun napping. He half rolled slowly and we got on his tail. Gosh, it's unpleasant fighting at that altitude. The slightest movement exhausts you."[5] As the aircraft clawed their way higher, the human flame began to gasp and sputter. Despite its superior mechanical performance, the Fokker lost its military advantage when thin air apparently snuffed out its pilot.[6] An untimely nap is just one symptom of hypoxia. Other signs include exhaustion, inattention, loss of voluntary muscle control, hearing impairment, hot flashes, and vision loss. Psychological effects range from fear and anger to laughter and euphoria.[7] These symptoms are sometimes too subtle to notice before a victim slips into unconsciousness and dies.

In a World War Two publication titled, "How to Avoid Death from Oxygen Failure," the Army Air Forces warned airmen that hypoxia is a treacherous, stealthy enemy. It seduces its victims. With "no violent pain-signal," airmen may not recognize symptoms until it's too late.[8] On 20 October 1943, for example, hypoxia claimed the lives of three of the ten airmen on a B-17 mission over Germany. Ten minutes from the target and cruising at 31,500 feet, the flight engineer discovered the left waist gunner curled up near his gun, his oxygen mask off; the right waist gunner was slumped over at his station, his oxygen mask on but its hose disconnected from the ship's supply; the radio operator sprawled prone in his compartment, his mask on but disconnected from its hose. In his attempt to resuscitate the radio operator, the engineer's oxygen hose detached from its source, and he fainted. The copilot found him in time, however, and the engineer survived. The others were dead.[9] In 1938 a prominent flight surgeon lamented that hypoxia is a quietly insidious pathology and not "an extremely painful process."[10]

Rapid improvements in aircraft performance in the 1930s exposed pilots to another altitude-related disorder that does involve extreme pain: decompression sickness (DCS), or the "bends." In 1670 Robert Boyle observed with his vacuum pump that reducing atmospheric pressure could cause bubbles to form in animals: "The little bubbles generated upon the absence of the air . . . [may] disturb or hinder the due circulation of the blood . . . [and] this production of bubbles reaches even to very minute parts of the body."[11] In one experiment Boyle even noticed "a conspicuous bubble moving to and fro in the waterish humour of [a viper's] eyes."[12] In the nineteenth century, divers and other compressed-air workers risked the bends when they emerged from their high-pressure environment into normal atmospheric conditions. The French scientist Paul Bert surmised in 1878 that decompression created bubbles of nitrogen in the blood; he established, moreover, that a *slow* rate of decompression and the breathing of pure oxygen limited bubble formation.[13] The advent of aviation provided a new venue for this phenomenon.

Rapid transition from high to low pressure caused nitrogen bubbles to form in joints, nerves, and blood vessels. Victims would writhe (or bend) in agony like the hapless creatures in Boyle's vacuum chamber. A swift ascent in an unpressurized aircraft could provide the same effect as that experienced by divers swimming too fast to the surface or by bridge builders emerging too quickly from their pressurized caissons. The formation of nitrogen bubbles—or *aeroembolisms* in the patois of aviation medicine—can result in DCS symptoms ranging from minor itching to debilitating pain.[14] Severe cases, such as Lieutenant Colonel Kevin Henry's U-2 mishap, include sequelae such as breathing difficulties, neurological impairment, or death.

The medical problems posed by hypoxia and DCS were not unique to airplanes. Acceleration-induced loss of consciousness, however, occurred only among aviators.[15] The disabling effects of acceleration forces were discovered when pilots initiated sharp turns or attempted to pull out of steep dives at high rates of speed. Flying a fast-moving aircraft in a tight turn or pulling out of a steep dive imposes acceleration forces, or G-forces, on the aircraft and the pilot. These forces are the same ones familiar to roller coaster enthusiasts experiencing sharp turns and loops. Unlike a roller coaster, air combat may subject pilots to prolonged G-forces. A sustained 5-G turn in an aerial dogfight or during recovery from a steep dive, for example, makes a 200-pound pilot weigh 1,000 pounds. He or she is pulled down into the seat; physical movement becomes difficult; and, most importantly, blood drains from the head. As the G-forces continue to pull a pilot's blood toward his feet, his vision fades toward total blackout, and he drops into unconsciousness until the G-forces are reduced.[16] The military disadvantages of this phenomenon are obvious.

As discussed later in this chapter, flight surgeons determined that pilots begin to lose their vision around three Gs and experience G-induced loss of consciousness, or G-LOC, around five Gs. These phenomena became more common as aircraft speed and maneuverability increased. In 1928, for example, the *New York Times* reported episodes of pilots momentarily fainting while performing aerobatic maneuvers.[17] A year later the *Aeronautic Review* referred to this disorder as the pilot experiencing a *blank*, an apt term for what happens when all of the blood rushes out of one's head.[18] By World War Two it became apparent that fighter aircraft could sustain more Gs than pilots could endure, unless the pilots were protected by some sort of technological enhancements. The P-51 Mustang, for example, could exceed seven Gs and promptly disable its pilot.[19]

The afflictions awaiting aviators were psychological as well as physiological. Consider some excerpts from a 1918 report from Britain's Royal Air Force cataloging various effects of combat aviation.[20] One pilot was granted leave because he "lost some of his dash," while another was "not so happy over the lines as he used to be." Another was "hopelessly stale," and a medical officer diagnosed one airman as "worked to death—wants to jump out of his machine." One flying officer was deemed temperamentally unfit when he could no longer peer over the side at 10,000 feet; another experienced insomnia and nightmares, collapsing when he saw enemy aircraft. Indeed, many of these airmen were as debilitated as those who suffered from hypoxia or the bends. Although the type and degree of psychological trauma defied precise measurement, flight surgeons worked throughout the interwar years to characterize what one medical officer labeled the "concussion of the mind."[21]

In 1920, US Army Air Service flight surgeons invoked the term *staleness* to describe nervous disorders that hampered an airman's efficiency.[22] This term competed in the 1930s with other labels, such as aviator's neurasthenia, flying stress, aeroneurosis, and flying fatigue. In World War Two, the Army Air Forces deployed psychiatrists to evaluate the effects of aerial combat and recommend treatment. The case of a B-17 tail gunner is instructive. During a training mission at 26,000 feet on 2 October 1942, one of the B-17's aircrew fainted from hypoxia. The pilot responded with a steep dive to 20,000 feet; while pulling out of the dive, the right wing snapped off, creating an explosion that severed the tail section from the fuselage. Trapped in the free-falling tail section, the tail gunner squirmed through an escape hatch and opened his parachute about 1,000 feet above the ground. He was the sole survivor. On a subsequent operational mission, his B-17 was raked with German gunfire, and he nearly bailed out without orders from the pilot. He later presented himself to the medical authorities. His complaint: fear of flying. He could not overcome his anxiety, nightmares, insomnia, and tremors, nor could he bring himself to enter an airplane. The Eighth Air Force Medical Board determined that no known therapy could return him to functional status and assigned him to ground duties.[23] The tail gunner's story illustrated that the hostile regime of flight—especially when combined with the risks of combat and accidents—bred disabling ailments in body *and* mind.

The Legitimation of Flight Medicine

A British aviator during the First World War remarked, "The ideal flier for our job is a youth of eighteen who has been crossed in love."[24] Flyers learned that aviation afforded little escape from the dangers of modern war. The death rate due to combat and accidents for US Army officer pilots in World War One was over five times the rate for officers in all other services.[25] More pilots died because of poor training or their own failings and shortcomings than from mechanical failure or enemy action. Early in the war nearly half of all flying cadets experienced a neurosis or nervous breakdown during training.[26] In 1914 90 percent of British aviation casualties resulted from "defects of the pilots themselves."[27] These statistics resulted, at least in part, from an initial lack of selection standards. Excerpts from medical records, for example, show that one British airman was blind in one eye. Another was denied entrance to the infantry three times and deemed a "very bad sailor." He was made a pilot.[28] In 1915, however, the British established a Care of the Flier Service that mandated higher physical and psychological standards. Greater medical supervision paid off for the British: fatalities due to "physical and mental defects" plummeted to 20 percent by the war's second year and 12 percent by the third.[29]

The US Army required its aviators to meet physical standards different from those of the average soldier. A 1912 War Department memo provided detailed instructions for physical exams: the candidate must demonstrate perfect vision, pass a color-blindness test, and be able to hop about with eyes open and shut.[30] One of the Army's first pilots, Lieutenant Benjamin Foulois, recognized in 1913 that pilots must not have a weak heart or weak lungs if they hoped to function at high altitude, and they must possess "strong nerves and well-trained muscles" to endure long flights and stormy weather.[31] The Army promoted aviation medicine when it appointed a chief surgeon to the Signal Corps Aviation Section in September 1917. At the same stroke, it created the Medical Research Board to investigate pilot health and develop experiments regarding oxygen deficiency.[32] The board created six separate medical departments: cardiology, physiology, psychology, otology, ophthalmology, and neurology and psychiatry.[33] It also established the Medical Research Laboratory in 1918. The laboratory's main charter was to conduct an eight-week course to train physicians regarding the physiological effects of flight, and it graduated the Army's first three "flight surgeons" in May 1918.[34]

Flight surgeons refined the aviation physical examination into a rigorous inspection of respiration, circulation, coordination, concentration, and psychology. They became gatekeepers of Army aviation, culling the unfit from the fit.[35] By 1918, aviation candidates had to do more than just pass a vision test and hop about with their eyes shut. Indeed, rumors about draconian tests flew through the ranks of American volunteers. Subjects heard of mallet-wielding physicians who tested the candidate's ability to regain consciousness after a sharp rap on the skull. Other tests were rumored to involve the use of a pistol fired behind the candidate to test his nerve. The *New York Times* declared such stories the work of German propaganda,[36] and a more scientific debunking of such stories appeared in a 1918 issue of the *Annals of Otology, Rhinology, and Laryngology*.[37] Even if Army flight surgeons resisted the temptations of mallets and pistols, they exerted significant authority over airmen. They decided who would become a pilot and determined whether he remained a pilot. A flight surgeon's professional judgment was enough to ground a pilot temporarily or permanently. Such power boosted the professional status of flight surgeons, but the increasing legitimacy of flight medicine must also be viewed in the larger context of early twentieth-century medicine.[38]

American medicine had transformed itself by the time of the Great War. New discoveries in biochemistry, physiology, and pathology permeated medical practice in the Progressive Era, and an objective, laboratory-based approach to diagnosis and treatment became routine.[39] New scientific perspectives were applied in the clinic through various forms of technology. Physicians relied on discoveries in chemistry to analyze blood and urine, and they harnessed Roentgen rays, or X-rays, to obtain images of the body.[40] Laboratory tests, quantitative measurements, and internal imagery complemented physician intuition. The application of science to medicine, especially when conducted through the medium of sophisticated gadgetry, enhanced professional prestige as well as a patient's trust.[41] Medicine had become scientized, and high-tech praxis enhanced legitimacy. Cardiologists in the early 1900s, for example, employed Willem Einthoven's electrocardiogram to identify abnormalities. This practice advanced the reputation of their specialty, even though little could be done to correct abnormal function.[42] Similarly, flight surgeons adopted the latest developments in science and technology to advance their art.

Between the world wars, flight surgeons and physiologists utilized various sophisticated scientific instruments to decipher how humans reacted to flight. These devices included altitude chambers to assess the effects of low pressure, centrifuges to test resistance to G-forces, electroencephalograms to evaluate epileptic tendencies and "emotional stability,"[43] and radioactive gases produced by Ernest Lawrence's cyclotron to analyze the movement of gas through the lungs.[44] The employment of such unique scientific instruments advanced flight medicine's legitimate role in the medical community and promoted the status of flight surgeons in Army aviation. Ultimately, flight surgeons translated new methods of representing and testing the body into tools of authority regarding pilot screening and performance standards.[45] The human material in aviation came under the province of flight surgeons, and improvements in aircrew performance relied on their technoscientific methods.

Since the pathology of flight involved physiological and cognitive issues, it is worth noting how the scientizing of medicine in the Progressive Era occurred in both physiology and psychology. Coincident with advances in other branches of medicine, psychology enhanced its professional status by adopting its own technoscientific approach.[46] This approach employed technical instruments to yield objective, quantifiable data. Indeed, the era between 1880 and 1910 was psychology's "brass age," an age when delicate instruments translated mental processes into quantitative form.[47] Psychologists wired their subjects to various instruments that recorded physical responses, such as reaction time, blood pressure, and pulse. Interrogation and introspection, the traditional modes for the study of the human mind, often proved incommensurate with such characterization and technoscientific methods.[48] Many psychologists, therefore, turned to behaviorism as a way to obtain new, measurable insights into the mind and to integrate scientific principles into their craft.

Behaviorism, with its reliance on modern instruments, lent an air of objectivity and scientific legitimacy. It became especially relevant in flight medicine since flight surgeons were concerned with how to predict and manage human response to various stimuli.[49] For example, in 1920 the Psychology Department of the Army Air Service's Medical Research Laboratory at Hazelhurst Field, New York, evaluated the speed and accuracy of typical pilot movements through measurements of reaction times. The trend toward objective measurement and analysis also appeared in the form of intelligence tests such as the ability to remember complicated movements or recognize familiar terrain.[50] Psychologists convinced the Army during World War One that formal IQ testing could provide "authoritative knowledge."[51] Similarly, flight surgeons developed their own technoscientific means of measuring and evaluating airmen in order to create authoritative knowledge and advance their professional standing.

Other means of legitimation for flight surgeons involved the development of their own educational and research institutions, the creation of special publications, and their fight for status as actual airmen. The Medical Research Laboratory, which included the new School for Flight Surgeons, studied flight-related physiology and psychology and applied the knowledge gained through this research in training physicians to be physiologists, clinicians, and psychologists of flight.[52] Established in 1918, it produced a steady stream of flight surgeons throughout the interwar years. By 1920 the school expanded its syllabus from two to three months. In 1922 the school changed its name to the School of Aviation Medicine (SAM) and in the same year graduated the first naval flight surgeon.[53] The Navy relied on the SAM as its source for flight surgeons until it built its own school in 1940.[54] In addition to institutionalizing flight medicine research and education, flight surgeons enhanced their professional status by issuing various publications.

Before establishing their own journal in 1930, flight surgeons relied on Army publications such as the *Air Service Information Circular* to report their research. A 1923 version of the circular, for example, contained thirteen articles whose topics ranged from cardiovascular changes induced by low oxygen to the problems of ocular adaptations for night flying.[55] In 1929 Louis H. Bauer, M.D., a former commandant of the School of Aviation Medicine, established the Aero Medical Association for civilian and military physicians interested in aviation medicine. A year later the association published the first issue of the *Journal of Aviation Medicine*. This monthly journal recorded many of the advances in flight medicine that took place in World War Two. Flight surgeons and researchers from civilian institutions such as the Mayo Clinic or the University of Michigan also published their findings in the *Journal of the Aeronautical Sciences*. In this journal, an article on aircraft structural design could be followed by an article on the structural limits of pilots.[56] The School of Aviation Medicine developed its own publication in 1937 called *Flight Surgeon Topics*. In one of its first volumes, the chief of the Air Corps' Medical Section declared that a flight surgeon "cannot be competent . . . unless he himself flies with the pilot, gets into the air and appreciates its problems and hazards, and by doing so is an airman himself, capable of giving advice which will produce results."[57] Thus, in addition to establishing their own professional publications, flight surgeons sought influence in Army aviation by attaining the status of flyers.

In 1920 the Medical Research Laboratory investigated a new way to improve the practice of aviation medicine. It sent one of its flight surgeons, Army captain F. C. Dockeray, through flight school so he could experience firsthand the lifestyle of his patients and the pathology of flight.[58] Dockeray completed the standard eight-week course and earned a Reserve Military Aviator rating. He learned to land his Curtiss JN-4 "Jenny" with a stalled engine, perform aerobatic maneuvers such as loops and spirals, fly in formation, and avoid tailspins.[59] Pilot status allowed him to live with other aviators and understand their stresses and ailments. Pilots spoke to him with greater candor, and he concluded that flight surgeons should fly in order to gain credibility with their patients. Although it was rare for flight surgeons to obtain an actual pilot rating in the decades after World War One, the SAM required physicians to log approximately fifty hours in the air observing aircrew activity before they qualified as flight surgeons.[60]

Flying provided flight surgeons an advantage in both substance and perception. Putting flight surgeons in the air enhanced their scientific judgment. The

SAM insisted that flying prepared a flight surgeon's mind for insights into pilot performance and enabled firsthand observations and appraisals of flight.[61] This also enabled flight surgeons to experience the environment in which their patients worked. This improved their bedside manner since they could relate to their patients. It made them better doctors. Flying also enhanced their reputation among airmen. In 1939 the chief of the Air Corps, General Henry "Hap" Arnold, believed that flying bolstered pilot trust of flight surgeons and argued, "If we keep the flight surgeons out of the air, we have reduced their value to us at least seventy percent."[62] The experience and credibility gained by active flight status permitted flight surgeons to collaborate with airmen and engineers to shape overall system performance.

Conclusion

By World War Two, flight surgeons held a respected position in military aviation. They set the physical and psychological standards for airmen; they exercised authority over the daily care of flyers; and they conducted innovative research to explore the pathological effects of flight and maximize human performance. Even Hollywood recognized the physiological dangers of flight and the vital role of aviation medicine. In 1941 Errol Flynn starred as a Navy flight surgeon in *Dive Bomber*, a film dramatizing the effort to "prevent the unconsciousness which comes at the end of a power dive [and] fight the strange and unpredictable ailments that attack a flying man high in the blue."[63] Flynn's character conducted scientific experiments in an altitude chamber, crafted technology to counter physiological threats, and enabled American pilots to fly higher than their Axis enemies. He even bested the pilots by getting the girl.

The professional status of aviation medicine was aided by a technoscientific ideal that produced new and authoritative knowledge. Moreover, flight surgeons established a legitimate medical specialty by creating their own institutional standards and publications. Their direct interaction with aviators not only boosted their credibility in the squadron but aided their research and clinical care as well. In essence, they became clinical pathologists of flight. Historian Stanley Reiser described clinical pathology as the investigation of the living. Pathology links science to medicine, and Reiser argued that new technologies of internal imagery, recording instruments, and physicochemical analysis enabled twentieth-century physicians to fuse laboratory research and clinical practice.[64] Flight surgeons did the same. They employed sophisticated technology to discover and characterize the pathological effects of flight. They observed airmen in situ and determined how humans endured the physiological insults of high altitude and high G-forces. Their work contributed to the survival of airmen and the optimization of military performance; at the same time, it demonstrated the limitations of human physiology on manned flight. Commensurate with their role as clinical pathologists of flight, flight surgeons also served the air force as engineers of the airman's body, and their work shaped the relationship between airman and aircraft.

Engineering the Human Machine

Interpreting the human body in mechanical terms has a long tradition. In 1628 William Harvey presented a mechanical explanation for the circulation of blood that viewed the heart as a pump. René Descartes, a contemporary of Harvey, surmised that mechanical philosophy could explain many physiological phenomena.[1] The French physician Julien La Mettrie argued in his 1748 text, *Man a Machine*, that mental and physiological phenomena resulted from complex mechanistic causes.[2] In the nineteenth century, Claude Bernard depicted organisms as living machines whose physicochemical processes constituted a *milieu interieur* discernible by scientific experimentation.[3] Physicians, according to Bernard, "must penetrate into the inmost phenomena of living machines and define their mechanism in its normal as well as its pathological state."[4] Such imagery would persist in aviation medicine.

Flight surgeons characterized the functions and limitations of the human machine in flight. They also crafted various means to mitigate the multiple pathologies of flight, and in doing so they took on the mantle of aeromedical engineers. In this role, they effected a new collaboration with aircraft designers, pilots, and air force officials and created an important nexus between life science and aviation. Indeed, flight surgeons wrote new prescriptions for aircraft designers: the unique properties and weaknesses of the human machine could not be ignored. By the mid-1930s the Army Air Corps institutionalized the vital relationship between aeronautical and aeromedical engineering. This established aviation medicine's key role in transforming the airman–aircraft relationship and shaping the evolution of flight.

Aeromedical Engineering

The head of the Army's Medical Research Board in 1918 described pilots as "human machines" and directed flight surgeons to maintain them at their peak efficiency.[5] This required flight surgeons to discover how humans functioned under various operating conditions, and they employed this knowledge to effect improvements in military aviation. If physicians functioned in a sense as engineers of the body, then flight surgeons were aeromedical engineers who sought ways to improve the performance of the human machine in flight. They were joined by other specialists to form a community of practitioners dedicated to aviation medicine. This community included physiologists, biophysicists, anthropologists, and others from the Army, Navy, private industry, and foreign countries. One way to interpret how their work shaped the evolving relationship between aircrew and aircraft is to employ concepts of technological change pioneered by historian of technology Thomas P. Hughes.

In his study of early electric power systems, Hughes introduced the idea of a *technological system* to describe the interconnection of varied components to achieve a common goal.[6] Power generators, distribution lines, human operators, appliances, and customers all served as vital elements of a complex whole. Similarly, when pilots climbed into the cockpit, they became part of a complex technological system. Overall system performance depended on their skills and limitations as well as on other elements, such as engine horsepower, aeronautical design, ground technicians, and life-support equipment. Hughes's system concept also incorporated an external perspective: systems were shaped by forces such as economic issues and institutional goals.[7] In comparable fashion, Army aviation was animated by technical, scientific, economic, organizational, and sociopolitical factors. For example, the Air Corps sought to expand the military potential of bombers by making them fly higher, faster, and farther.[8] This emphasis, however, imposed significant strains on a pilot-centric technological system. Improvements in aeronautical capabilities advanced system performance but only to a point: pilots soon became a limiting factor and presented what Hughes described as *reverse salients*.

In Hughes's model of technological systems, problems with individual components can limit the capability of the entire system. In Thomas Edison's nascent electric power system, for instance, direct current to transmit long distances proved uneconomical; this constituted a reverse salient since it impeded overall performance of the power generation and distribution system. Only the introduction of alternating current allowed the entire system to advance.[9] Reverse salients, in other words, are weaknesses in one or more components of a technological system that impede progress in the entire system. Hughes coined the term as an analogue to a military campaign: as a front advances, some parts of it may encounter resistance that stalls forward progress, and this forms a reverse salient in the advancing line.[10] In a similar manner, as military aviation advanced in terms of aircraft performance, it encountered an array of reverse salients related to pilot limitations. These included the altitude-related disorders of hypoxia and decompression sickness. In order for system performance to advance, flight surgeons engineered solutions to mitigate these problems.

Oxygenating the Flyer

While training the initial class of flight surgeons in 1918, the Army installed American aviation's first altitude chamber at the Medical Research Laboratory in Mineola, New York. This device, the largest in the world at the time, could accommodate up to six human subjects.[11] It was essentially an updated version of Robert Boyle's seventeenth-century vacuum pump. Instead of small animals, flight surgeons sealed larger subjects—pilots and oftentimes other flight surgeons—into a chamber and pumped out an alarming percentage of the air. They then recorded the effects of this new environment that simulated high altitude. A major emphasis was on categorizing airmen in terms of their ability to tolerate reduced oxygen levels. Chamber experiments revealed that respiration and cardiac output increased commensurate with a drop in air pressure, and the ability to function differed among individuals. Some adapted quickly, others did not.[12] In the absence of reliable oxygen equipment, the Army restricted flyers who performed poorly in low-oxygen tests to altitudes below 15,000 feet. Some aviators were limited to 8,000 feet, and an unlucky few earned a "should not fly at all" rating.[13] According to the American Expeditionary Force Commission on Flight Surgeons, "It would be murder to allow a man to fight a Hun at 18,000 feet" if he was unfit to do so. It was "a distinctly American idea," the commission added, "to develop a system for maintaining at the highest possible level of effort the human machine, without which the most perfect airplane was worthless."[14]

One method of examining how the human machine operated at high altitude was to integrate various devices with the altitude chamber. In 1918 the Medical Research Laboratory rigged its altitude chamber with an X-ray apparatus. Technicians replaced one of the chamber's windows with an aluminum panel, so that subjects in the chamber sat with their backs pressed against the aluminum with a photographic plate suspended in front of them. An X-ray outside of the chamber shot radiation through the aluminum to produce images of how low pressure and hypoxia affected cardiac morphology and function.[15] The laboratory also incorporated a reaction-time apparatus into the chamber to evaluate cognitive changes induced by low pressure. This permitted flight surgeons to characterize the mental as well as physical impact of high altitude. The chamber also had the capability to reduce ambient temperature to forty degrees below zero in order to determine the combined effects of cold and altitude.[16] Extreme cold simulated the realities of flight above 20,000 feet, and creating this environment on the ground enabled flight surgeons to make improvements to aircrew oxygen equipment that was otherwise prone to icing up. The use of high-tech instruments cohered with the technoscientific ideal in aviation medicine and aided flight surgeons' efforts to overcome pilot limitations.

Armed with increasingly sophisticated knowledge of hypoxia, flight surgeons sought to engineer reliable, comfortable oxygen equipment that could sustain pilots at high altitudes. During the First World War, aviators used a crude pipe stem to suck in some occasional oxygen; the pipe stem was replaced by several versions of rudimentary masks, most of them uncomfortable and susceptible to leaking or freezing. Such unreliable oxygen equipment enervated the human component and generated a reverse salient. For example, when Army Air Service major R. W. Schroeder attempted to break the world altitude record of 34,910 feet in 1920, his aircraft was equipped with the latest performance improvements crafted by the Air Service's Engineering Division, including an improved propeller, a supercharged carburetor, and pressurized fuel lines. But the modified airplane lacked suitable means to sustain its pilot. Schroeder distrusted his crude oxygen mask, so he relied on sticking a rubber hose in his mouth. At about 33,000 feet, well below his aircraft's projected altitude capability, Schroeder fainted due to hypoxia. Witnesses observed his aircraft plummet out of control nearly six miles. Three thousand feet above the ground Schroeder regained his senses and landed safely. He later reported: "My next sensation after losing consciousness was a terrific explosion in my head, and on looking about, I discovered I was within a few thousand feet of the earth."[17] Clearly, much work remained for flight surgeons in terms of integrating man and machine.

Although flight engineers developed some functional improvements in oxygen systems, as late as 1940 a dependable oxygen mask proved elusive. Moreover, supplemental oxygen equipment produced its own drawbacks: it added excess weight to aircraft, required careful maintenance, reduced mobility for aircrew in large aircraft such as B-17 bombers, and hindered communication.[18] Nevertheless, military imperatives pushed aircraft to altitudes that required new devices to maintain human function and enable overall system advancement.

On 8 May 1929, a US Navy pilot wearing a customized, form-fitting oxygen mask set a new altitude record of 40,000 feet. Despite improvements in his oxygen equipment, the pilot experienced weakness, dizziness, and disorientation.[19] He hoped to attain at least 42,000, but his aircraft entered a spin as it climbed through 40,000 feet. The spin may have saved his life since it caused a rapid descent to thicker air. Five years earlier, altitude chamber tests had suggested the absolute ceiling of human tolerance occurred between 40,000 and 45,000 feet, even *with* the availability of oxygen from a tightly fitted mask.[20] Had he flown higher, the pure oxygen in his lungs would have been at too low of a pressure to dissolve into the bloodstream—he would have suffocated even while breathing pure oxygen. In World War Two, flight surgeons helped engineer a new technology, pressure breathing, to overcome this reverse salient. Pressure breathing narrowed the widening gap between human limitations and aeronautical advances.

Pressure breathing, or positive pressure respiration, involved the inhalation of air forced into the trachea under high pressure. It was uncomfortable and required active, forceful exhalation. Pressure breathing served as a clinical therapy for patients with lung-related ailments such as bronchitis, asthma, and pneumonia several years before flight surgeons adapted it to aviation.[21] In November 1941 Captain John Kearby, an engineer from the Equipment Section of the Army Air Forces Materiel Command at Wright Field, asked flight medicine researchers if it was possible to "supercharge" the lungs. Kearby knew that supercharged engines could achieve higher altitudes, so he figured something similar might work to increase performance in the human machine.[22] Kearby's notion intrigued an Army biophysicist, Colonel A. P. Gagge. Gagge crafted a pressure breathing apparatus and tested it on himself in the altitude chamber. The device enabled him to reach 46,000 feet with little detriment, a significant improvement over standard oxygen equipment.

Further research established that pressure breathing could raise the performance ceiling for the average pilot from 40,000 feet to about 43,000 feet. The Army Air Forces employed pressure breathing equipment for high-altitude reconnaissance flights over Europe and Japan in 1944.[23] To prevent the lungs from dangerous expansion during pressure breathing, researchers developed a corset-like device that squeezed the pilot's torso.[24] In short, airmen were being engineered, at considerable discomfort, to enhance their ability to keep up with aircraft capabilities. Humans were being adapted to their machines.

Pressurizing the Flyer

While developing a scientific understanding of hypoxia and engineering the means to prevent it, flight surgeons also turned their attention to characterizing and mitigating the threat of decompression sickness. This malady became familiar as advances in aeronautical engineering enabled the rapid ascent of fighter aircraft and the prolonged high-altitude cruise of bombers. It presented yet another reverse salient in aviation advancement and was the subject of numerous studies in the late 1930s and 1940s. A flight surgeon noted in 1941 that aircraft performance had exceeded a pilot's ability to adapt: "Twenty years ago the prospect of developing aeroembolism or bends because an airplane could ascend to an altitude at sufficient speed to develop these symptoms was unheard of. Today it is a very live subject."[25]

In World War Two, Army Air Forces airmen had to endure a trip to the altitude chamber before they could participate in modern air war. They were locked inside a steel chamber with a heavy door, most of the air was sucked out, and they met their deadly opponent of high altitude. All of them learned to recognize and alleviate the symptoms of hypoxia, and in their fight to remain conscious many also experienced decompression sickness. Of the 68,422 trainees who endured the altitude chamber at Maxwell Field, Alabama, during World War Two, more than 9,500 experienced some form of the bends. Researchers at Maxwell noticed that a one-hour flight at 38,000 feet produced aeroembolism in 27.4 percent of trainees, and half that number got sick during the same period at 30,000 feet.[26] Notably, the ceiling of B-17 Flying Fortresses and their fighter escorts extended into this altitude range. Flight surgeons also discovered that age, excess body fat, and above-average height all correlated with a higher incidence of the bends. Race, they noted, was not a factor.[27] The tests demonstrated that decompression sickness became likely during combat at high altitudes in unpressurized bombers. In order to decrease this risk, aeromedical engineers experimented with pressure suits.

The Air Corps began tinkering with pressure suit development in 1934. Engineers at the Army's aeronautical development center at Wright Field, Ohio, collaborated with a civilian test pilot, Wiley Post, to conduct the world's first flight in a pressure suit. Post, an eccentric character, was one of the foremost test pilots in the interwar years. In 1933, for example, he set a record by making a seven-day solo flight around the world in his own airplane, the *Winnie May*.[28] He also designed his own high-altitude pressure suit. The suit was a rubber garment made by the B. F. Goodrich Company with an outer layer of tough fabric. The helmet, affixed to the suit by metal strips and wing nuts, amounted to little more than a large aluminum can with a window. The window was small since Post had only one eye (his other eye having been lost in an oil-drilling accident). Post inflated the suit with pressurized oxygen from a liquid oxygen tank.[29] It was a far cry from the sophisticated apparatus worn seventy years later by Lieutenant Colonel Kevin Henry in his U-2, but it worked during altitude chamber tests at Wright Field. On 28 August 1934, Post climbed into the *Winnie May* and made his record flight. Although he attained altitudes above 40,000 feet on multiple test flights, the *Winnie Mae* lacked the engine power to break the altitude record of 47,352 feet.[30]

After testing Wiley Post's outfit, the Air Corps continued to experiment and issued a call for prototype suits in 1939. B. F. Goodrich, Goodyear Tire and Rubber, and the United States Rubber Company all attempted to develop reliable, convenient pressure suits for sustained high-altitude operations.[31] By 1942 numerous tests were performed with these suits in the altitude chamber and in actual bomber aircraft. The tests were only a partial success: the suits worked but the aircrew could not. B-17 test flights in 1942 proved particularly instructive. General Grandison Gardner, the commander of the Army Air Forces Proving Ground Command at Eglin Field, Florida, volunteered as one of several test subjects. Wearing only a shirt and shorts, he donned his rubber suit at 5,000 feet and turned a dial on the suit to fill it with air and create a personal pressurized environment. Although the suit became bulky and somewhat rigid when inflated, General Gardner managed to operate the waist gun and the radio station. He also climbed into the cockpit and flew the bomber through various maneuvers. The general complained of difficulty bending his arms and legs, and he thought the helmet should be connected lower on the neck to improve flexibility and visibility; although he was able to pencil a note while in full regalia, he recommended improving the gloves as well.[32] The general's problems were minor compared to those encountered in other tests.

The pressure suit's immobilization of its occupant proved to be its biggest liability in the B-17 tests. Most of the test subjects gave less favorable reviews than General Gardner. The helmet and rubber garment were too heavy for prolonged missions, and when filled with pressurized air, the apparatus became too bulky for efficient movement. Encased in their suits, airmen waddled around, slow and off-balance, protected yet immobile, like knights weighted down with cumbersome armor. Once costumed in their giant rubber prophylactics, pilots complained they couldn't "feel" or finesse their airplane. Bombardiers could not sit at their station and lean forward far enough to manipulate their controls; instead, they had to kneel on the floor behind their seat and contort themselves to perform required movements.[33] The Army Air Forces Medical Laboratory declared the apparatus "air-worthy but, in its present form, not combat-worthy."[34]

Although pressure suits were designed to prevent decompression sickness, they also held promise as a protection against acceleration forces. Applying pressure around the legs and torso slowed the drainage of blood from the head during sharp turns at high speed. Flight surgeons thus explored the dual potential of pressure suits to prevent both aeroembolism and G-induced loss of consciousness. This concept caught the attention of Vannevar Bush, the chairman of the National Advisory Committee for Aeronautics (NACA). In October 1940 Bush wrote to the chief of the Army Air Corps, General Henry "Hap" Arnold, urging the development of pressure suits for pilots. Bush declared that the problems of aeroembolism and G-induced loss of consciousness involved "a small amount of physics and aerodynamics, and a great deal of physiology, and probably of psychology."[35] A month later the Army responded to the NACA chairman, assuring him that researchers were investigating the physiological benefits of pressure suits.[36] Indeed, in the decades after the Great War it became apparent that improvements in military aviation required "a great deal of physiology." As flight surgeons gathered scientific data on the internal workings of the human machine, they also served as aeromedical engineers and devised improvements in the man–machine interface. These improvements advanced system performance by overcoming reverse salients associated with high-altitude flight. In the course of their work, however, they discovered that acceleration forces imposed an ultimate physiological limit to the potential of manned aircraft.

Keeping Flyers Conscious

In his 1925 essay on aircraft evolution, Army Air Service major Henry "Hap" Arnold predicted that aircraft speed and maneuverability would reach an absolute limit dictated by the capabilities of human operators.[37] His prediction came true. Improvements in aircraft performance during the interwar years revealed that acceleration forces imposed a physiological threat utterly unique to high-performance aviation. In 1927, for example, US Marine Corps pilots became more aware of acceleration effects as they worked on bombing techniques against insurgents in Nicaragua, and US Navy pilots experienced similar problems as they developed dive-bombing procedures in the 1930s.[38] By the time Major Arnold became a five-star general in World War Two, some types of fighter aircraft could pull more Gs than their pilots could endure. System performance became limited by the human machine.[39] This particular pathology of flight posed what historian of technology Edward Constant has theorized as a *presumptive anomaly.*

In his analysis of the development of jet engines, Edward Constant described how the inherent aerodynamic properties of propeller technology limited aircraft to subsonic speeds. Eventually, propeller-driven aircraft would achieve a maximum level of performance. A final but subsonic speed record would be set due to propeller limitations despite sleek new aerodynamic improvements that would otherwise permit higher speeds. Constant called this condition a presumptive anomaly; progress in a certain line of technological development would be capped by a theoretical upper limit.[40] As aeronautical engineers improved the speed, altitude, and maneuverability of combat aircraft, pilots became the limiting factor in the technological system. Human physiology would limit aircraft maneuverability regardless of improvements in speed. Constant argued that a presumptive anomaly is "science-based" and recognized by "those very close to the intellectual foundations of their respective fields."[41] Flight surgeons played this role.

Flight surgeons simulated the physiological effects of G-forces in the laboratory using a human centrifuge. In 1935 flight surgeon Harry G. Armstrong constructed the Army's first centrifuge at the Air Corps Materiel Division, the center of research and development in Army aviation.[42] Suitable for human and animal subjects, Armstrong's "acceleration machine" used an electric motor at the center of a twenty-foot beam. Test subjects were placed in a compartment at either end of the beam and whirled around to simulate the G-forces experienced during sharp turns in flight. This device allowed flight surgeons to evaluate the physiological effects of forces up to thirty Gs.[43] They hooked subjects up to a variety of recording instruments: oximeters to measure blood oxygen levels, sphygmomanometers to measure blood pressure, and other devices to record pulse and respiration. The results helped describe the symptoms experienced by airmen during aggressive aerial maneuvers and offered a scientific basis for the presumptive limits in maneuverability of manned aircraft.[44]

The centrifuge tests revealed how human performance ebbs as acceleration forces exceed three Gs. Flight surgeons observed that around three Gs pilots experienced a graying of their vision; they lost peripheral sight and their vision narrowed to a small point, as if peering through a straw. As acceleration forces increased from four Gs to six Gs, pilots experienced a rapid transition from grayout to blackout to complete loss of consciousness. At five Gs or higher they were reduced to blindness and then unconsciousness in as little as four seconds.[45] Their heart still pounded, they still drew breath, yet no blood reached the brain. Remaining at high Gs would be fatal, but once a pilot fainted during flight he would unconsciously release the controls and the acceleration forces would diminish. Blood could then flow back to the brain. After experiencing a G-induced loss of consciousness, the average pilot remained *hors de combat* for about thirty to sixty seconds. Upon regaining consciousness, however, he could find himself in a worrisome position: either stalled, spinning, and pointing at the ground, or locked in the enemy's gunsights. Armed with an improving scientific understanding of how acceleration forces disrupted human physiology, researchers investigated several means to counteract this disorder.[46]

Borrowing from work done by the US Navy, along with efforts by Canada and Australia, US Army flight surgeons used the human centrifuge to develop an "anti-blackout" suit, or G-suit, during World War Two.[47] Worn like a pair of pants with a large waistband, the G-suit contained numerous air bladders that inflated during the onset of acceleration forces, constricting around the legs and abdomen and stemming the drainage of blood toward the feet.[48] Centrifuge testing demonstrated that the G-suit could increase pilot tolerance by an average 1.2 G. Moreover, flight surgeons invented a procedure that a fighter pilot could perform to improve his ability to remain conscious under high G.

With the goal of increasing blood pressure in the upper torso, a pilot learned the art of tensing the muscles of his legs and abdomen, taking in quick, sharp breaths, and producing a loud grunt as he forcefully exhaled every few seconds. This inconvenient maneuver, familiar to those suffering from constipation, worked. For the average pilot, such tensing and grunting added yet another one G of resistance.[49] These innovations offered substantial advantages during aerial combat. Japanese pilots sought a similar technological edge by binding their legs with inelastic fabric to the point of extreme discomfort.[50] Flight surgeons also determined that pilots could add another 1.2 Gs of tolerance by leaning forward from an upright position into a crouch.[51] In the high-G environment, therefore, military necessity transformed the swashbuckling fighter pilot into a grunting, bent-over machine operator sporting painfully tight pants.

Despite research on the effects of acceleration forces in the mid-1930s, the Germans failed to develop their own G-suit. The director of aviation medicine research at the Army Air Forces' Materiel Command noted after the war that the Luftwaffe was "amazed that our P-51s could maintain a superior ratio of fighter kills. . . . They attributed this to our 'G-suit' or anti-blackout costume, which enabled our airmen to out-maneuver the faster enemy aircraft."[52] The Germans had taken a different approach, modifying their aircraft instead of their pilots. The Luftwaffe employed an automatic pullout device for the Junkers Ju 87 "Stuka" dive-bomber. This technology took over when the pilot released a bomb during a steep dive and enabled the aircraft to recover to normal 1-G flight, even if the 6-G pullout rendered the pilot unconscious.[53] Automating the aircraft to compensate for the inadequacies of the pilot suggested a means to bypass the presumptive limit imposed by the pilot. Moreover, it offered a glimpse of a technological trajectory toward automation and pilotless airplanes (discussed in later chapters). Before such advances, however, Army flight surgeons sought other means to adapt the human machine to the high-G environment.

Blood pressure could be increased only to a certain limit before risking damage to the circulatory system.[54] Flight surgeons, therefore, pursued alternative methods to manipulate the body in order to improve aircraft performance. One innovative technique sought to transform the way pilots flew: instead of flying in a seated position, they would lie prone, much like Orville Wright at Kitty Hawk in 1903. Orville lay prone to operate a wing-warping mechanism with his hips, and this also reduced wind resistance. A few decades later the Germans tested whether the prone position held promise, especially regarding resistance to acceleration forces.[55] By 1939 they illustrated that a prone pilot could endure fourteen to seventeen Gs without physiological insult, a vast improvement over the four to six Gs in the sitting position.[56] In 1943 the Division of Medical Sciences of the National Research Council determined that prone pilots could endure at least nine Gs for prolonged periods without visual problems.[57] The Army Air Forces pursued prone flying during the Second World War in order to increase aircraft maneuverability, but with mixed results.

Flight surgeons and other life scientists at Wright Field evaluated the prone position in terms of comfort, ability to access and operate controls, and visibility. While suggesting that aircraft could be flown from the prone position, they determined that such a change would generate a parade of problems. Pilots would require, for example, a special sling under the chin that suspended their head and allowed them to maintain a forward gaze.[58] One investigation after the war noted that in order for pilots to move their head around while lying prone during high-G conditions they would require a special helmet connected by cables to counterweights; otherwise, they became pinned down and unable to look about during high-G maneuvers. The position was also complicated by bulky oxygen masks.[59] Moreover, it required new methods of restraint. A special harness would have to hold the body in place, and this generated safety concerns regarding crashes or ejections.[60] Laying prone also reduced rearward visibility and created a blind spot vulnerable to attack from behind. Overall, flying while lying down posed more problems for pilots than if they remained seated.

In addition to altering the position of the pilot within the aircraft, aeromedical engineers sought chemical means to manipulate human physiology and expand the theoretical limits of human performance. Detlev Bronk, chief of the Division of Aviation Medicine in the Office of Scientific Research and Development during World War Two, noted that physiologists tested a variety of compounds to improve human performance. Adrenocortical hormone and ammonium chloride, for example, showed some promise in boosting resistance to hypoxia. Other substances such as adrenalin, Benzedrine, and ephedrine were evaluated for their ability to increase tolerance to acceleration forces.[61] Yet the pharmaceutical approach proved disappointing. The short-term benefits were marginal and unpredictable, and the long-term side effects were undefined.[62] Pharmaceutically enhanced flying, therefore, never reached operational status except in the form of pep pills that kept airmen alert on long missions.

Even if the prone position or pharmaceutical intervention had been successful, the presumptive anomaly of human G tolerance would have remained. At some point, after overcoming various G-related reverse salients, aircraft maneuverability would become limited by human G tolerances, even if the pilot laid prone, used stimulants, or both. This undermined the notion of pilot supremacy, a notion already battered by human susceptibility to hypoxia and decompression sickness. By the mid-1930s, human frailties highlighted the need for a revised approach to aviation development. Cruising at high altitude posed significant physiological problems for bomber crews in particular. Fighter pilots risked the same difficulties when they ventured up to high altitudes, but they also faced G-related impairments during rapid, high-speed maneuvers in aerial dogfights and dive-bombing. Progress in the overall technological system of military aviation would require both a new perspective on the pilot's role and an institutionalized interface between aeronautical and aeromedical engineers.

Interface between Aeronautical and Aeromedical Engineering

Aided by physiologists and biophysicists, flight surgeons functioned as engineers of the airman's body. With their knowledge of the pathological effects of flight, they advanced the technological system by fashioning means to improve human performance. Flight surgeons treated the pilot as "the heart and brain of the whole flying apparatus" and set about improving the function of heart, brain, lungs, eyes, and ears during the rigors of flight.[63] They viewed the airman as the most vulnerable and important component in the flying system, and the tendency of aircraft engineers to ignore biological factors was a constant irritant. As early as 1919, for example, the Army's Office of the Surgeon General in the Division of Military Aeronautics lodged this complaint: "Wonderful has been the development of the airplane—inconceivable has been the neglect of the MAN in the airplane."[64] Flight surgeon Harry G. Armstrong echoed this sentiment in a 1936 *New York Times* interview: "They build planes almost as planes alone. They tell us that 'this one has a ceiling of 30,000 feet,' but they have yet to ascertain what effect such high flying will have on the crew that will fly that plane daily, or upon the passengers."[65]

The disconnect between those who designed aircraft and those who flew them reflected what historian I. B. Holley termed "design hubris."[66] This concept described the failure of designers to consider various important elements in the technological system. Detlev Bronk recognized design hubris as the development of technology that operators cannot use without extreme discomfort or danger.[67] The B-17 Flying Fortress provided a fitting illustration. It flew high enough to subject its crew to bitter cold, hypoxia, and decompression sickness yet offered little protection against such hazards. If a machine gun jammed, the gunner had to remove his bulky gloves and risk freezing his hands to the cold metal, all the while breathing through an unreliable oxygen mask.[68]

Engineers designed the B-17 to reach high altitudes in order to reduce vulnerability to enemy countermeasures, but in their zeal they overlooked biological issues and thus compromised the overall system. Holley equated this disconnect with an arrogance born of ignorance. Still, it would be unfair to accuse early aircraft designers and engineers of completely ignoring the human element.

The Army's *Handbook of Instructions for Airplane Designers* gave at least fleeting recognition of the human component. The 1921 version, for example, directed engineers to avoid obstructing the pilot's vision. The 1934 version provided more guidance: designers should allow adequate space for airmen to egress the aircraft with a parachute, and the seat and rudder pedals should be adjustable to accommodate variations in pilot height. Even after thirteen years, though, the *Handbook* gave scant emphasis to basic biological requirements such as heat and oxygen. Still, many engineers in the interwar years recognized the importance of pilot opinions and preferences. Aircraft designers in the Army and researchers at the National Advisory Committee for Aeronautics sought input from pilots regarding the flying qualities or "feel" of aircraft, and they devised changes to aerodynamic control systems based on pilot feedback.[69] Engineers also sought the opinion of flight surgeons when working with Wiley Post to design early pressure suits. Yet despite these examples, aeronautical engineers interacted rarely with life scientists until the mid-1930s.

One way to limit design hubris in the development of technological systems was to develop organizations that blended different communities of practitioners.[70] This occurred in 1935 when the Air Corps created the Physiological Research Unit in the Experimental Engineering Section of the Materiel Division at Wright Field, Ohio. This institutionalized a new relationship between life scientist and aeronautical engineer, which in turn shaped the relationship between airman and aircraft.

The formation of the Physiological Research Unit was precipitated by the actions of Major Malcolm C. Grow and Captain Harry G. Armstrong, flight surgeons in the Army Air Corps. Grow joined Wright Field's Experimental Engineering Section in 1931. His purpose was to serve as flight surgeon to the section's engineer-pilots. Some of these pilots had already achieved notoriety in aviation.[71] Lieutenant Albert Hegenberger made the first flight from the United States to Hawaii in 1927, and Lieutenant James Doolittle conducted the first zero-visibility flight and landing in 1929. Hegenberger and Doolittle worked with several other aviators in a wide variety of endeavors. Indeed, Grow served in a hothouse environment of aviation research and development. This was the site for wind tunnel experiments, parachute trials, Wiley Post's pressure suit tests, advances in instrument flight, construction and appraisal of aircraft engines and armament, and high-altitude balloon studies.[72] In the course of his duties as flight surgeon, Grow conducted modest, informal physiological research and offered his medical perspective to some of the official research projects in the Experimental Engineering Section. In 1934 Major Grow left the section for Washington, DC, to serve as the chief flight surgeon of the Air Corps.[73] This positioned him to aid the endeavors of his replacement, Captain Harry G. Armstrong.

Captain Armstrong's medical expertise proved particularly valuable to the Experimental Engineering Section's *Explorer I* high-altitude balloon project. A three-man team of pilot-engineers—Major William Kepner, Captain Albert Stevens, and Captain Orvil Anderson—collaborated with the National Geographic Society in 1934 to explore the upper reaches of the atmosphere and attempt to break the altitude record of 61,237 feet set in 1933. They planned to remain safely tucked away in a sealed, air-tight gondola and operate a variety of scientific instruments, including a Geiger counter to measure cosmic rays. Armstrong assisted in the testing of their pressure and oxygen supply and noted the dual importance of gathering information about the stratosphere as well as the medical issues associated with it.[74] Despite careful preparations to protect the human passengers, the *Explorer I* mission failed in a spectacular manner.

As the aviators ascended through 57,000 feet, a rip appeared in the balloon fabric. This slowed the climb, and the flight's altitude peaked just 624 feet shy of the record. At that point more rips developed, and the *Explorer I* entered a harrowing descent. Kepner, Anderson, and Stevens remained helplessly sealed in the gondola until they returned to thicker air below. When their dangerous fall continued below 4,000 feet above the ground, they decided to abandon ship. Kepner and Anderson got out first and parachuted uneventfully to safety. But the balloon exploded just as Anderson leapt clear of the gondola. Stevens had trouble getting through the escape hatch after the explosion. On his third attempt, as the gondola streaked toward the ground, he cleared the hatch and managed to parachute safely to earth even though some of the balloon debris became entangled with his parachute canopy.[75]

Undeterred, and still confident in their life-support system, Stevens and Anderson launched the *Explorer II* mission the following year and shattered the altitude record by two miles. After his experience with the *Explorer* flights, Armstrong wrote a seminal account of "the medical problems of sealed high-altitude aircraft compartments" in 1936.[76] This paper laid the foundation for future specifications of pressurized aircraft. It also paved the way toward a new organizational entity, the Physiological Research Unit.

Captain Armstrong approached the chief of the Experimental Engineering Section, Major Oliver P. Echols, concerning the need for an official, funded physiological research capability within the section. Echols agreed and submitted a proposal to the chief of the Materiel Division, Brigadier General A. W. Robins, on 25 April 1935.[77] In the meantime, Major Grow worked his connections in the higher echelons of the Air Corps. Grow met with General Robins and Colonel F. L. Martin, a staff officer for the chief of the Air Corps, and argued the merits of a physiological research laboratory at Wright Field.[78] On 6 May 1935 General Robins responded to Major Echols's proposal and Major Grow's unofficial appeals and sanctioned the establishment of a Physiological Research Unit with Armstrong at the helm.[79] Robins's imprimatur acknowledged the importance of medical research to aviation development and the "efficiency, health, and lives" of airmen. Aeromedical research gained an even greater patron that year when Brigadier General Oscar Westover, the assistant chief of the Air Corps, ordered the Physiological Research Unit to enable flying personnel to "function efficiently under the adverse and often abnormal conditions experienced in flight."[80]

Patronage from the highest levels of the Air Corps enabled an official connection between life scientists and aircraft engineers. By 1937 construction was completed and the Physiological Research Unit took a new name, the Physiological Research Laboratory. Although the laboratory pursued various projects ranging from more efficient oxygen masks to measurements of oxygen saturation in the blood during high-altitude flight, Armstrong focused its efforts on the development of pressurized aircraft.[81] This endeavor highlighted the importance of incorporating biological considerations into aircraft development.

The concept of cabin pressurization was simple: maintain an artificially high level of air pressure within an aircraft as it flew at high altitude. "Supercharging" the cabin with pressurized air would negate the need for oxygen masks and pressure suits as well as maintain comfortable cabin temperature. Operationalizing this concept proved difficult and required engineers and life scientists to resolve a host of problems. Indeed, fourteen years before establishing the Physiological Research Laboratory, the air force dabbled with the concept of pressurized flight. This first effort failed to account for numerous technical considerations, though, including basic physiological needs. It was a fiasco.

In the summer of 1921 the Army Air Service adapted a British De Havilland DH-9 biplane bomber by removing the seats for the pilot and observer and inserting a large steel tank that could be filled with pressurized air.[82] Just big enough to hold one pilot, the tank sported a twenty-four-inch-wide door and five six-inch portholes, and it was rigged with the basic controls necessary for flight. Cockpit instruments such as the airspeed indicator could be viewed by the pilot through one of the portholes. A propeller-driven compressor pumped pressurized air into the tank, and the pilot could adjust the air pressure by opening a pressure-relief valve, or so the engineers thought. On its maiden flight, the DH-9 climbed to 3,000 feet; the pilot, Harold R. Harris, then locked the steel door in place, imprisoning himself inside the cockpit/tank. The compressor pumped air into the cockpit/tank at an alarmingly high rate, driving the atmospheric pressure to the equivalent of 3,000 feet below sea level. The compressor was more efficient than expected; even with the relief valve wide open, the pressure continued to increase. Meanwhile, the temperature rocketed up to 150°F.[83] It was little more than a flying pressure cooker. Harris tried to open the door, but the pressure held it shut. He scoured the cockpit/tank for something to break one of the windows. Nothing was available. He considered cutting off the engine to reduce the pressure, but remembered the compressor was driven by a separate propeller mounted on the wing and would continue to pump in air regardless of engine power. Peering through the portholes, Harris managed to navigate back to the runway and land before succumbing to the intense heat. Apparently, the project's funding dried up after this incident, although Harris remembers that engineers planned to equip future flights with a cutoff valve for the pressure intake and, just in case, a hammer.[84]

The 1921 episode portended some of the physiological and technical challenges of pressurization. On the basis of the *Explorer* balloon flights, Armstrong established in 1936 the critical physiological requirements of carbon dioxide removal, temperature regulation, barometric pressure management, and oxygen supply.[85] One year earlier an engineer at the California Institute of Technology outlined some of the major design problems created by pressurizing a cabin. In addition to the severe weight penalty, these problems included developing a reliable automatic system to control air pressure, ways to divert engine power to air compressors, standard requirements for door and window designs, and sufficient space for auxiliary equipment such as valves and piping.[86] Pressurized aircraft, therefore, presented an immense medical and technical challenge yet also held great promise for the ambitions of airpower enthusiasts. In 1935, one month before establishing Armstrong's Physiological Research Unit, the Air Corps Materiel Command initiated the development of an experimental aircraft, the XC-35, to explore the military efficacy of pressurized flight.[87]

The Air Corps contracted with Lockheed to construct the XC-35. Lockheed converted the fuselage of a two-engine, ten-passenger Electra into a sealed tank by installing heavier doors, smaller windows, and seals around the aircraft's rivets and seams.[88] The Air Corps named a Berkeley engineer, Professor John E. Younger, to lead the project at Wright Field, and Younger assigned Armstrong and the new Physiological Research Unit the task of researching relevant physiological issues.[89] Armstrong's laboratory and Younger's engineers made a good team since the success of the project depended on the guidance of both flight surgeon and engineer. Their cooperative effort overcame a variety of developmental problems.

Unlike the ill-fated 1921 attempt at pressurized flight, the XC-35 engineers decided to test the pressurization system on the ground with no one inside. It was a good idea since the first hangar test turned into something of a circus. As the pressure in the fuselage increased, the air-tight seals began to crack, and the aircraft started to emit a chorus of whistles. As the pressure rose, the whistles became screeches, and onlookers began to grow wary of this shrieking calliope. They made a hasty retreat toward the hangar's exits as the XC-35's windows began to bow outward.[90] Technicians fixed the seal problems, but the experience alerted Armstrong to the effects of a rapid decompression, or sudden loss of pressure at high altitude. What would happen to the aircrew if a window blew out?

In a 1936 paper on pressurized compartments, Armstrong contemplated the medical effects of rapid decompression. He surmised that a pilot would experience ruptured ear drums, stabbing sinus pain, and a "displacement of the heart and lungs with resultant embarrassment of their action." The pilot could also expect the rapid onset of hypoxia, and his blood would froth with evolved nitrogen gases. Armstrong also predicted physical and psychological shock.[91]

If he was right, flight at high altitude in a pressurized cabin could place airmen at extreme risk. Experiments in the Physiological Research Unit allayed some of these fears. The unit conducted animal tests to study the effects of rapid decompression and determined that airmen could survive, given appropriate countermeasures. They would not go into shock, feel severe pain, or experience heart or lung failure; instead, they simply had to don an emergency oxygen mask and descend in order to avoid hypoxia or decompression sickness.[92] Engineers, therefore, would have to provide for auxiliary oxygen equipment in case of depressurization.

The first pressurized flight of the XC-35 took place on 25 May 1937. Its occupants could now cruise at high altitudes without wearing bulky winter gear or oxygen masks. A War Department press release noted the value of such technology to commercial travel as well as military operations.[93] The XC-35 became a flying laboratory in which engineers tested pressurization equipment and flight surgeons studied physiological matters. The XC-35 project received the 1937 Collier Trophy, an annual award granted for the most significant advance in aviation.[94] Its success demonstrated the value of close cooperation between aeromedical and aeronautical experts. This disciplinary alliance between flight medicine and aeronautical engineering evolved beyond the confines of Wright Field during the late 1930s and World War Two.

Another formal interface between life scientists and aeronautical engineers materialized in 1940 in the form of the Committee on Aviation Medicine in the National Defense Research Council.[95] In 1941 the Committee on Medical Research (CMR) in the Office of Scientific Research and Development subsumed the Committee on Aviation Medicine. These organizations helped bridge the gap between physiology and aeronautics by coordinating aviation medicine research with aircraft production. They directed some of the aeromedical research noted earlier in this chapter, including studies of oxygen equipment, pharmaceutical enhancements of human performance, and methods to improve night vision.[96]

In 1948 Detlev Bronk, chief of the CMR's Division of Aviation Medicine during World War Two, attested to the importance of the aeromedical–aeronautical interface. While he condemned physicists and engineers for their design hubris in the two decades before Pearl Harbor when they created machines that unnecessarily imperiled the lives of their operators, he claimed the designers had earned their redemption when they accepted aviation medicine as the key to advancing the technological system: "We have attained our present prowess in the air through the combined efforts of the physical scientist and the physiologist."[97]

Conclusion

The pathology of flight shaped technological change in military aviation. The scientific characterization of afflictions such as hypoxia, decompression sickness, and G-induced loss of consciousness enabled flight surgeons to engineer means to improve human performance. By the mid-1930s their efforts to advance the technological system of manned flight were aided by organizational changes that merged aeromedical and aeronautical expertise. These changes reflected the common view in aviation medicine that flight surgeons must "write the prescription of physical requirements before the engineers could proceed with their designing."[98] Advances in the technological system of military aviation came to rely, therefore, on an increasingly complex interaction between life science, aircraft engineering, and military requirements.

This interaction was symbolized in a 1941 exchange between Major Harry Armstrong and General Henry "Hap" Arnold, the chief of the Air Corps. Seven months before Pearl Harbor, Armstrong visited England for a flight surgeon's perspective on the high-altitude operations of Spitfire and Hurricane pilots. Improvements in engine technology had enabled these British fighters to increase their combat ceiling above 30,000 feet. Armstrong noted that pilot efficiency deteriorated between 20,000 and 25,000 feet, and pilots became careless and lethargic above 30,000 feet, even while wearing oxygen masks. He concluded that the British fighters should be equipped with pressurized cockpits "without delay."[99] General Arnold commented on this report two days later. He had spoken with Armstrong and concluded that "it is absolutely essential" for American pilots to be trained to fight at altitudes up to 37,000 feet in unpressurized aircraft until "supercharged cabins or pressure suits or some similar arrangement is made available."[100] Arnold's decree declared a military requirement for flight operations in environmental extremes. It also acknowledged that modern air war required more than efficient wings and powerful engines; the pilots themselves would have to be supercharged, or somehow integrated with their machine, in order to exploit the air weapon.

Aviation medicine balanced a scientific understanding of human capabilities with technical efforts to maximize system performance, yet its prescriptions stunted aircraft potential. Even the most efficient integration of the human component impeded machine performance. Manned aircraft could operate at high altitudes for extended periods, but only by burdening the aircraft with heavy life-support equipment. Regardless of the efficacy of aeromedical interventions, aircraft maneuverability was limited by human physiological capacity. Aeromedical engineers, nevertheless, continued to seek ways to improve the symbiosis between airman and aircraft. Their work shaped aviation evolution and conformed to the Flight Surgeon's Oath, in which they committed themselves to "the healing of the mind as well as the body." To achieve that goal, they vowed: "I will be ingenious. . . . I will be resourceful . . . [and] strive to do the impossible."[101]

Flying Blind

Because pilots were unable to function in some flight regimes without technological enhancements, the military imperative to fly in environmental extremes elevated the status of flight surgeons and engineers and altered the relationship between airmen and aircraft. Humans had to adapt to their machines. Similar concepts applied in regard to the development of instrument-based flight. While other histories of instrument flight supply a useful narrative of how technological advances were incorporated into military and commercial aviation,[1] the analysis in this chapter considers how physiological limitations sparked fundamental innovations that fashioned a different relationship between humans and their flying machines.

Just as low barometric pressures and high acceleration forces impaired pilot function, blind flight, or flight in low-visibility conditions, rendered pilots incapable of maintaining their spatial orientation. Fog, clouds, and darkness not only hampered the ability to navigate but often resulted in loss of basic aircraft control. Pilots, flight surgeons, and engineers all sought new ways to cope with these hazards. In 1926 a flight surgeon and a pilot collaborated in a seminal experiment. They used a gyroscopically powered turn indicator to demonstrate that pilots became hopelessly disoriented during blind-flight conditions not because of inattention, inexperience, or incompetence; rather, disorientation and its resultant, often fatal mishaps were simply due to *normal* human physiology. In the late 1920s and 1930s, the military and commercial aviation communities developed special instrument technology to compensate for physiological limitations and enable humans to operate in blind-flight conditions. By World War Two, aviators were indoctrinated into a new paradigm of flight that replaced their own instinctive senses and judgment with instrument-based control of aircraft. These developments involved a complex interplay of physiology, physics, engineering, and military and commercial demands, and they helped establish the trajectory toward an increasingly complex technological system that redefined the role of airmen.

The Inability to Fly Blind

In 1932 Winston Churchill warned, "The air is an extremely dangerous, jealous, and exacting mistress."[2] He spoke from personal experience. He learned how to fly when he was First Lord of the Admiralty from 1911 to 1915, and his diplomatic shuttles as air minister after World War One often found him at the controls. His love affair with flight was occasionally a dangerous one: he survived an engine failure, a crash during takeoff that destroyed his aircraft, a stall after takeoff that destroyed yet another aircraft, and a fire in the fuselage extinguished in the nick of time. Despite such hazards, Churchill recognized that even gifted pilots "perished too often in some ordinary, commonplace flight."[3] The air became especially dangerous and exacting when pilots encountered the commonplace conditions of fog, clouds, or darkness. Churchill cited an example from 1914 when one of his flying buddies, a skilled pilot "three parts bird and the rest genius," disappeared forever in the twilight fog over the English Channel.[4] Limited-visibility conditions became a common nemesis that claimed the lives of experienced and inexperienced pilots alike and led to US Army aviators to coin the term *blind flight* when flying in conditions where one could no longer see the ground.

The long catalog of aviation accidents suggests that pilots lacked any inherent talent for blind flight. It was not uncommon for a pilot on a routine sortie to fly into clouds or fog, lose sight of the ground, become disoriented, and emerge from the fog bank or cloud in a spin, stall, steep spiral, or even upside down.[5] Major General Mason Patrick, the chief of the Army Air Service from 1921 to 1927, told of a blind-flying incident where the pilot realized he was upside down only when his watch dropped out of his pocket and struck him in the face.[6] Lieutenant Elmer Rogers, an Army Air Corps flight instructor from 1928 to 1933, illustrated the dangerous, disorienting effects of flight in limited-visibility conditions with his account of a disastrous 1929 sortie flown by "Lieutenant X."

While flying from San Antonio to Brownsville, Texas, a formation of five Air Corps biplanes encountered a fog bank. Lieutenant X maintained his position at the rear of the formation as the lead aircraft descended to tree-top level in order to skim just below the fog. Avoiding ground obstacles while remaining out of the fog required quick, abrupt maneuvers, and at one point Lieutenant X pulled into a steep ascent to avoid collision with a wingman. He was suddenly in the fog, no longer able to see the other aircraft or the ground. He could reduce power and descend, trying to feel his way toward the terrain. Or he could continue to climb and break into the clear. Choosing the latter course, Lieutenant X advanced the throttle and waited to ascend into sunlight. He never did.

As Lieutenant X attempted to climb out of the fog, he noticed several unanticipated indications on his aircraft's basic instruments. The compass had begun to spin. The airspeed read 110 miles an hour—too fast, especially for a climb. The altimeter indicated a steady decrease rather than an increase in altitude. He suspected that the aircraft had somehow nosed down, so he pulled back on the stick to raise the nose, but this only increased his speed. Alarmed and confused by this state of affairs, he tried a combination of inputs on the control stick, yet nothing worked. The compass still spun, the airspeed still increased, and the altitude still decreased. At this point Lieutenant X decided to abandon the aircraft. This posed a dilemma since he had a passenger, an enlisted photographer. The lieutenant motioned the photographer to jump overboard but was met with a bewildered, frightened stare. Unable to see the ground, and failing to compel his passenger to jump, Lieutenant X leapt from the aircraft and pulled his ripcord. Just as his parachute opened he heard the aircraft impact the terrain. The passenger died.[7]

Untrained in how to interpret and trust his instruments, Lieutenant X had no ability to discern his relationship to the horizon. In his confusion, he was unable to interpret his rotating compass and increasing airspeed as indications of a spiral toward the ground. Notably, Lieutenant Colonel Kevin Henry experienced a similar disorientation nearly eighty years later when he flew his U-2 into the clouds and, unable to interpret his instruments, entered a steep, descending spiral. For Lieutenant X, pulling back on the stick tightened the downward spiral instead of raising the nose as he expected. His panicked, undisciplined inputs only aggravated the scenario. Flying out of the spiral and into a wings-level climb would have required a precise, coordinated application of controls, a simple maneuver on a clear day. Lieutenant Rogers instructed his students that performing the same maneuver in a fog bank presented an impossible task without the aid of special instruments.

Lieutenant X deserved some credit for realizing something was wrong. Other pilots who survived similar situations thought they were still in "ordinary commonplace flight" and never recognized that their aircraft was out of control. Trusting in their own instincts and physical sensations, they assumed that their airspeed, altitude, and turn indicators had malfunctioned shortly after entering the weather. Lieutenant Rogers noted that many pilots convinced themselves that clouds were magnetic since not long after entering a cloud the magnetic compass tended to rotate.[8] These supposed instrument malfunctions, of course, reflected the disorientation of the pilot and the erratic flight path of the aircraft rather than some sort of coincidence or weather-induced glitch. They also exposed the dangers of pilot hubris.

Lieutenant Rogers told this story to warn his students about the manifold treacheries of blind flight. As aviation evolved during World War One and the 1920s, most aviators realized that blind flying increased the risk of disorientation and loss of aircraft control. Still, many believed that a pilot's sense of balance and "seat-of-the-pants" instinct enabled flight through clouds or fog, and failure to fly blind suggested a lack of skill. Lieutenant Rogers, on the other hand, argued that even gifted, experienced pilots lacked the ability to remain oriented to the ground during blind flight.[9] As discussed later, his beliefs were based on a new scientific understanding of blind flight developed by a flight surgeon and an Army aviator in 1926. Lieutenant Rogers instructed his students that, in order to avoid the same fate as Lieutenant X, they must learn to distrust their own senses and rely instead on special cockpit instruments. When interpreted properly, instruments that indicated airspeed, altitude, and direction of turn could provide a truer representation of reality than a pilot's instinctive "feelings."

The Necessity to Fly Blind

Blind flight posed an obstacle to effective navigation as well as a danger to airmen. Even if pilots could maintain their spatial orientation during flight in limited visibility, finding their destination required additional instruments that portrayed the aircraft's physical location relative to known landmarks. Because the problems of unaided blind flight limited flight to clear weather and hobbled the potential of military and commercial aviation, special instruments that oriented the pilot to the surface of the Earth became increasingly important as aircraft improved their range and duration.

A 1923 observation by Lieutenant Albert F. Hegenberger, an Army Air Service pilot and a forerunner in the development of blind-flying technology in the interwar years, illustrated an early recognition of the connection between military efficacy and instrument flight. Lieutenant Hegenberger believed the air weapon's inherent mobility provided "vision and striking power," yet these advantages were meaningless without the ability to reach distant locations in poor weather.[10] Mobility required special cockpit instruments to fly in fog, clouds, or darkness. It also required precise navigation instruments for pilots to stay on course above an overcast, over trackless terrain, or away from coastal areas. Without such technology, all that remained was a fair-weather, short-range flying machine whose capability would hamper the commercial potential of aircraft and undermine the vision and efforts of airpower enthusiasts to exploit the air weapon.

After World War One, an initial priority of the air force was to use airpower to protect American shores by sinking ships far out at sea. Brigadier General William "Billy" Mitchell, an early airpower theorist and firebrand advocate for an independent air force court-martialed in 1925 for insubordination, demonstrated the efficacy of land-based bombers against a German battleship, the *Ostfriesland*, sinking it seventy miles off the Virginia coast in 1921.[11] This bolstered the efforts of airpower advocates to acquire the coastal defense mission, but it also illuminated the need for instruments that allowed aircraft to sally forth from their coastal bases, find their targets, and return in inclement weather.[12] General Mitchell experienced this firsthand when, after observing the *Ostfriesland*'s demise from his own aircraft, he encountered poor weather on the return to base and had to land in an open field miles from his destination. Coastal defense thus required some means to fly and navigate in blind-flight conditions. So did another vital mission: strategic bombing.

During the 1930s the air force emphasized the putative merits of long-range strategic bombing. Airpower enthusiasts envisioned armadas of bombers flying over entrenched armies and striking deep into enemy territory, crippling the adversary's war machine at its source. Central to this strategy was the concept of high-altitude daylight precision bombing. Air Corps strategists assumed that bombers could find and destroy distant targets on a reliable basis, yet they assumed too much and overlooked weather concerns. A 1939 memo from the General Headquarters (GHQ) Air Force to the chief of the Air Corps revealed the problem of striking precision targets in poor visibility.[13] The GHQ's operational tests showed that bombing within or above an overcast could be "fairly effective" only against large targets such as ports or cities, and the GHQ expressed doubt about striking precision targets hidden beneath clouds. To bomb a precision target under such conditions required flying below the overcast and risked "extravagant losses" to enemy defenses. Because such conclusions undermined the core strategy of American airpower, the GHQ urged immediate action to "increase the effectiveness of the bombardment airplane under all conditions by all practicable means."[14] This required greater reliance on instruments and development of effective navigation capabilities. The complex system of aerial bombardment, therefore, was only as good as the aircrew's ability to operate in all weather conditions.

Another instrument-related impediment to airpower employment involved aircrew training. Blind flight required careful instruction in the use of instruments, and the Air Corps struggled to equip its pilots with adequate skills. As new blind-flight instrument technology appeared in aircraft, airmen tended to lag behind in their ability to use it. In early 1941 Colonel Ira Eaker, the commander of the 20th Pursuit Group who became a leading advocate of high-altitude precision bombing in World War Two, complained to his superior that the group's fighter pilots were receiving inadequate instrument training even though "we have at least one foot in a war." Eaker's pilots needed more practice in blind flying, and he declared they were "useless" if they couldn't climb through the weather to intercept enemy bombers or escort friendly ones.[15] The Air Corps responded to Eaker's concerns and shipped twenty "hoods" to the group for blind-flight instruction. These hoods covered the cockpit, blocking the pilot's view of the outside and simulating the conditions of blind flight. This technique helped pilots develop proficiency in instrument flight, but the Air Corps cautioned Eaker that the hoods could cause accidents when used at low altitudes.[16] Instrument training, therefore, was not without its risks, and Air Corps leaders had to incorporate the costs of such training in the development of the air weapon.

Aircraft needed to operate in all types of weather conditions in order to function as a reliable implement of war or, for that matter, commerce. Hegenberger's view that improved performance required "new and more refined mechanisms" recalls the notion of reverse salients presented in the previous chapter.[17] Just as improvements in aircraft performance required special life-support technology, operations in blind-flight conditions required special instrument technology. In both cases, system advancement relied on new forms of technology to surmount human incapacities. Overcoming the reverse salients of pilot disorientation and inadequate navigation became just as critical to the development of airpower as overcoming the problems of hypoxia, decompression sickness, and G-induced loss of consciousness. The solution to both problems, moreover, involved advances in the science of human physiology and a retooling of the airman-aircraft complex.

The Physiology of Blind Flight

Humans maintain their balance and spatial orientation through a variety of sensory inputs. The children's game where one tries to remain balanced while spinning around blindfolded illustrates the importance of vision to the perception of motion and position. When not being spun around, however, blind people can remain perfectly balanced due to other sensory organs. Special nerves in the joints, muscles, and tendons called proprioceptors inform the central nervous system of the body's position. Other types of nerves in the skin serve as mechanoreceptors and sense touch. In addition to these faculties, the inner ear contains a vestibular system, a complex of tiny mechanisms that sense acceleration and rotation of the body. Although a blind man can remain oriented while on the ground thanks to proprioceptors, mechanoreceptors, and the vestibular apparatus of the inner ear, a pilot conducting blind flight becomes hopelessly disoriented due to the normal function of these senses.

The basic motions of an aircraft are pitch, roll, and yaw. Pitch is the amount of nose-up or nose-down movement around the lateral axis of the aircraft (the lateral axis parallels the wings); roll is the amount of turn around the horizontal axis of the aircraft (the horizontal axis parallels the fuselage); and yaw represents how much the nose swings to the left or right along the vertical axis of the aircraft (the vertical axis parallels the tail). Similarly, the vestibular system in the inner ear senses the pitch, roll, and yaw of the body with a tiny but elaborate mechanism resembling a three-dimensional carpenter's level. Three small, semicircular canals in the inner ear are arranged at right angles to each other and correspond to the lateral, horizontal, and vertical axes of movement. Each canal contains a viscous substance called endolymph fluid. As the head pitches forward or backward, rolls left or right, or yaws from side to side, special sensors in the equivalent semicircular canal detect the movement of the endolymph fluid within the canal. These sensors translate the mechanical effect of fluid movement into a neural impulse and allow the brain to interpret the body's movement.

One of the first insights into vestibular function came in 1825 when French physiologist Jean Pierre Flourens disabled the semicircular canals of pigeons. Destroying the horizontal canal disrupted the pigeon's sensation of yaw, and it flew in circles. Without its vertical canal, the pigeon lost its sense of pitch and tumbled about in somersaults. In 1861 a French otologist named Prosper Ménière surmised that vertigo resulted from a disease process in the semicir-

cular canals. Building on the work of Flourens and Ménière, Ernst Mach and others in the final quarter of the nineteenth century established the semicircular canals as the sensory organs of rotation.[18]

In the early 1900s, the Viennese otologist Róbert Bárány improved medicine's knowledge of the vestibular apparatus when he linked disturbance of the fluid in the semicircular canals with the sensation of bodily movement through space. In his *experimentum crucis,* Bárány elicited the perception of movement by injecting hot or cold water into the ear; the temperature change roiled the vestibular fluid and induced an overpowering sensation of movement in the patient. His work advanced the pathophysiological knowledge of the inner ear and earned him the 1914 Nobel Prize in Medicine.[19] The contributions of Bárány and his predecessors did not escape the notice of physicians interested in flight medicine. Even before the advent of the flight surgeon specialty, they assumed that pilots required a good sense of balance to maintain their equilibrium in the air. To ensure this capability, they employed Bárány's procedure to screen aviation candidates during World War One.

In addition to syringing hot or cold water into the ear, examiners used a "Bárány chair" that rotated its occupant at various speeds to induce vertigo. Either test created a dizzying effect in normal applicants. When a blindfolded candidate was spun around in the Bárány chair, for example, he would gradually lose the sensation of turning once the sensors in the appropriate semicircular canal became acclimated to the movement of the endolymph fluid. At this point, the examiner would stop the chair and whisk away the blindfold. A normal person would then exhibit rather peculiar behavior: his eyes would twitch back and forth, he would be incapable of pointing his finger at a stationary object, and he would immediately fall sideways, convinced he was now spinning in the opposite direction.[20] Ironically, such erratic behavior was precisely what examiners looked for: it indicated a healthy candidate's response to rapid, unpredictable alterations in the movement of the endolymph fluid in the semicircular canal. Dizzy and disoriented by the spinning and auditory douching, he was fit to fly.

Although aviators and flight surgeons considered a normal sense of balance an important prerequisite for flight, both communities misunderstood the nature of disorientation during blind flight. The director of the Air Corps' Instrument and Navigation Unit in 1937 noted that during World War One and until the mid-1920s, "there was supposed to exist a very select group of bad-weather pilots, mostly self-appointed, who were looked upon as being endowed with some special quality that made them able to control their airplane by their senses alone during flight without outside visual reference."[21] A common assumption was that proprioceptors, mechanoreceptors, and especially the vestibular system of the inner ear helped pilots resist the disorienting effects of blind flight. Thus, healthy pilots should be able to fly blind with enough practice. In 1917 an Army physician declared: "Without functionating [*sic*] internal ears it is impossible for an individual to be a good bird man. . . . Many an aviator has gone to his death because, all unknown to him, he did not possess a normal ear mechanism."[22] This was absolutely wrong. In reality, many an aviator went to his death because he *did* possess a normal ear mechanism. Consider again the unfortunate case of Lieutenant X.

Lieutenant X's testimony suggested that he allowed the aircraft to enter a turn during his attempted climb out of the fog. As he continued in the turn, the sensors in his corresponding semicircular canal acclimated to the turning motion and caused him to perceive that he was in level flight. When he pulled back on the stick to climb, he only tightened the turn toward the ground. If one assumes Lieutenant X was in a left turn, had he stopped the turn, the endolymph fluid in the semicircular canal would have shifted in the opposite direction and convinced him he was now turning to the right. In that case, his natural tendency would be to compensate for the right-turning sensation and turn back toward the left, thus continuing the leftward spiral—a *graveyard spiral* in aviation patois. Such spirals were little more than a trick of the inner ear. Without instruments, Lieutenant X's only options were to descend into clear weather (this is how Lieutenant Colonel Henry recovered from the downward spiral of his U-2), parachute out of the aircraft, or crash.

The misunderstanding about the relationship between the vestibular system and spatial disorientation during blind flight was illustrated not only by frequent weather-related mishaps but by attempts to desensitize pilots to the sensations of vertigo. Pilots endured devices such as the Ruggles Orientator, a crude simulator that rotated its occupants in all three axes in order to develop "immunity" to the disorienting effects of blind flight.[23] One flight surgeon during World War One, Major Robert J. Hunter, sought to improve pilots' sense of balance and ability to fly in poor weather by having them fly while wearing a blindfold. Hunter postulated that flying blindfolded would help pilots with their "feel of the ship." He planned a training program where pilots flew blindfolded every day for ten minutes; at the end of two weeks they were supposed to have developed an ability "to keep an even keel without the aid of eyesight."[24] Thankfully, the war ended before Hunter initiated this ill-conceived program.

The widely held notion that a pilot could develop immunity or resistance to vertigo implied that disorientation during blind flight reflected a disease or another pathological effect of flight. Many pilots were prone to declare themselves immune or resistant to dizziness or vertigo. In their view, such afflictions were experienced only by the novice, the weak, or those less endowed in natural ability. By 1926, however, it became clear that no one could develop any sort of immunity or resistance to the disorienting effects of blind flight. The realization that normal physiological function rendered humans incapable of blind flight catalyzed a fundamental shift in the nature of flight: instruments would not supplement human senses; *they would replace them*. Examining this pivotal change in aviation first requires an understanding of early instrument development.

The Purpose and Progress of Early Instruments

Instruments that aided the pilot's perception of the environment appeared in the early years of powered flight. The 1908 version of the Wright Flyer utilized history's cheapest and simplest flight instrument: a piece of string. Attached to the front end of the aircraft, it indicated if the pilot applied too much yaw to the left or right during a turn.[25] If the fluttering string aligned itself with the fuselage of the airplane, then the pilot knew he was applying the controls properly. Benjamin Foulois, one of the first Army aviators and a future chief of the Air Corps, learned to fly in 1908 and upgraded the Wright Flyer's string to a piece of tape about two feet long. According to Foulois, who characterized flight in those days as "no more risky than dodging Moro knives in the Philippines," the swinging of the tape to the left, right, up, or down served as a good indicator of the aircraft's attitude.[26] A British flyer named W. H. D. Acland utilized a similar principle but with considerably more fashion and flare. In his flying days before World War One, Acland would acquire a ribbon from "the prettiest girl in the neighborhood of the flying field" and tie it to the front strut. In addition to enhancing his social life, it served multiple purposes as a yaw or sideslip indicator, an airspeed indicator, and, while on the ground, a wind meter. The ribbon also possessed the virtues of being "crash proof, extremely light, needing no calibration, and very cheap."[27]

Strings, tape, and ribbons soon gave way to much more delicate, heavy, and expensive instruments as aviation evolved. By 1914, the Wrights added a rate-of-climb indicator as well as a pendulum-based device to indicate bank.[28] Throughout World War One various other instruments appeared in cockpits. Some aircraft were equipped with a Creagh-Osborne compass, a crude device placed on the cockpit floor and nestled in a horse hair cushion to minimize vibration.[29] Additional war-era equipment included instruments that measured airspeed, the degree of drift over the ground, the angle of bank, and altitude.[30] By 1917 the Signal Corps specified that its aircraft contain a magnetic compass and engine instruments such as oil pressure and gas gauges.[31] These instruments allowed a pilot "to do more intelligent flying."[32] Maintaining a straight course or monitoring one's airspeed and altitude in low-visibility conditions, for example, required the supplemental information provided by a compass, airspeed indicator, and altimeter. Despite these advances, the Signal Corps' organizational successor, the US Army Air Service, contended in 1919 that instruments "can never replace the sense of equilibrium and control—the 'feel' of the ship—which every pilot must have."[33] This flawed view conformed to the prevailing wisdom that a normal vestibular system aided rather than obstructed a pilot's ability to operate aircraft.

Although the fact that a pilot's vestibular system rendered him unable to fly blind was ignored until 1926, the aviation industry sought new ways to enhance pilot precision in inclement weather. In 1918 the Sperry Gyroscope Company produced a gyroscopically stabilized turn indicator. Elmer Sperry, a pioneer of gyroscope-based devices, described the turn indicator as a "crutch" since it would provide a more precise indication of the direction of turn than the pilot's senses or the magnetic compass, both of which tended toward instability and inaccuracy.[34] Aware of his interest in blind flight, the Sperry Company supplied a turn indicator in 1918 to Army Lieutenant William C. Ocker. Using this device, Lieutenant Ocker made a successful flight in the fog from Washington, DC, to New Philadelphia, Ohio, that year.[35] Eight years later, he would use this turn indicator to refute the conventional wisdom about blind flight and spatial disorientation. During those years, Army aviation continued its efforts to overcome the problems imposed by weather and darkness.

In the early 1920s, the Air Service conducted numerous operational tests of new instruments and navigational equipment. In 1923 an Air Service aircraft flew from Dayton, Ohio, to Boston; the remarkable aspect of this flight was that the pilots could not see the ground during 450 miles of the 750-mile sortie. They used a new device, the Earth inductor compass, to maintain their course.

Invented by the Bureau of Standards in 1920, this mechanism employed Michael Faraday's concept of electromagnetic induction by using the Earth's magnetic field and a rotating armature to produce an electric current that corresponded with the aircraft's heading.[36] This allowed the pilot or navigator to dial in the desired magnetic heading, and a needle would remain centered as long as the aircraft maintained this heading. The device garnered a favorable review during another Army test flight from Dayton to Seattle in 1925. The inductor compass functioned "exceedingly well" despite one significant mishap: when attempting to cross over mountains in eastern Oregon the aircraft nearly struck a mountainside because, in the pilot's words, "when the situation was most tense, [my] scarf became detached and wrapping around the impeller of the compass, caused the bronze horizontal jackshaft to break."[37] One can only imagine this frantic scene, and it suggested that much progress remained in the art and science of instrumented flight.

Further progress in the Air Service's effort to improve blind flight came in the form of the "flight indicator." This mechanism combined the gyroscopically driven turn indicator mentioned previously with a fore-and-aft inclinometer that indicated the degree of nose-up or nose-down tilt. Lieutenant Albert Hegenberger noted in 1923 that the flight indicator helped bomber aircraft maintain more precise flight parameters during their target run.[38] Instruments such as the flight indicator, altimeter, and compass provided only a rough approximation of the aircraft's position and direction, yet with extensive practice they could dramatically improve system performance. The three Air Service pilots who conducted an around-the-world flight in 1924, for example, were dedicated to the idea that blind flight and navigation required knowledge and expertise in the use of instruments rather than complete reliance on human senses.[39]

The Air Service seemed to realize the importance of instrument use by 1925. A Technical Regulation stated that pilots must depend on their airspeed, altitude, turn, and heading indicators; otherwise, they could "lose all sense of balance, become lost, and possibly fall into a spin."[40] Despite this injunction and the instrument-based advances in blind flight and navigation in the early 1920s, basic instrument training remained inadequate, and the Air Service failed to accelerate the incorporation of new instruments into its aircraft. By 1925, most aircraft contained outdated "war stock" instruments left over from the Great War. One report from the Air Service's Engineering Division at McCook Field, Ohio, noted that aircraft often lacked a proper compass, or when it was supplied, the compass was sometimes attached to a wing strut instead of a useful spot in the cockpit.[41]

Although instruments such as Sperry's turn indicator and the Earth inductor compass represented significant advances in instrument flight, inadequate training in instrument use and a general misunderstanding of spatial disorientation hindered improvements in all-weather aviation. In 1923 Brigadier General William Mitchell, the assistant chief of the Air Service, ordered the Engineering Division at McCook Field to equip some airplane cockpits with hoods so pilots could simulate the conditions of blind flight and improve their all-weather proficiency. The test was terminated, however, after the pilots proved unable to fly in a hooded cockpit due to physical reactions of vertigo and spatial disorientation.[42] This behavior posed an anomaly to the normative view that a skilled, healthy pilot with a good sense of balance could learn to operate in limited-visibility conditions. An explanation for this anomaly emerged from a chance interaction between a flight surgeon and a test pilot in 1926. Their subsequent collaboration transformed the relationship between airman and aircraft and redefined the nature of flight.

The "Ocker Vertigo Stopper Box" and a New Paradigm of Flight

Major David Myers served as an Air Service flight surgeon during the mid-1920s at Crissy Field in San Francisco. He believed that young aviators tended toward cockiness and a dangerous overconfidence in their abilities. A pilot's inflated ego presented an inviting target for Myers, who sought to "partially deflate it for his own safety."[43] Myers's tool of choice was the Bárány chair. As described earlier, the revolving Bárány chair allowed physicians to induce vertigo by stimulating the subject's vestibular system. Myers spun pilots around and around, allowing them to keep their eyes open. When asked which way they were turning they responded quite naturally with the correct answer. He then blindfolded them and repeated the maneuver; after about thirty seconds of rotating, Myers allowed the chair to come to a gentle stop. When asked what was happening, pilots answered that they were still turning, but in the *opposite* direction of the original turn. When Myers removed the blindfold, the scales fell from their eyes, and they could now see their own susceptibility to illusion. The young pilots left the exam less cocksure and more cautious, or so Myers hoped.

Myers was on the verge of a major insight. The inability of pilots to judge rotation in the absence of visual references explained the failure of General William Mitchell's training experiment with hooded cockpits in 1923. It also presented an anomaly to the conventional notion that a pilot's skill and ability, once properly developed, enabled safe passage through limited-visibility conditions. Myers tested this assumption by asking "an old-time pilot," Major William C. Ocker, to strap into the Bárány chair.[44]

Major Ocker was a lucky choice. He had enlisted twenty-eight years earlier during the Spanish-American War and served in a variety of posts as infantryman, aircraft mechanic, and, eventually, aviator. While a sergeant at Fort Myer, Virginia, in 1909, Ocker witnessed the Army's first test flights and became fascinated with aviation. In 1912 he requested transfer to the Aeronautical Section of the Signal Corps and completed training as an aircraft mechanic. Taking private lessons on the side, Ocker earned his pilot license in 1914 and was allowed to serve as a pilot and mechanic on Army aircraft. Ocker gained experience in a variety of aircraft over the next few years, and he received a Reserve commission in 1917.[45] In 1918 he flight-tested Elmer Sperry's turn indicator, and he carried it with him as an aid during flight in limited-visibility conditions. Ocker's suspicion that the turn indicator tended to malfunction in bad weather suggested that he, too, failed to recognize the natural tendency toward vertigo and spatial disorientation during blind flight.[46]

When Major Myers applied the same Bárány chair treatment to Major Ocker in January 1926, Ocker, like the younger pilots, and despite being one of the most experienced aviators in the Air Service, found his sense of orientation undone by rotational movement without visual cues. Even as a practiced aviator, he could not maintain his situational awareness when Myers simulated the conditions of blind flight. He left Myers's office with an idea. He returned for another Bárány chair session, this time equipped with the gyroscopic turn indicator acquired from Elmer Sperry years earlier. Ocker fastened the turn indicator, a compass, and a flashlight within a wooden box. With a hole cut in one end, he could peer into the box and observe the instruments while blocking his view of the outside world. Ocker spun up the gyroscope, and Myers spun the chair. Ocker then proceeded to ace the test. He gave the correct answer for every combination of turns, stops, and reversals by trusting the instruments within the box rather than his own vestibular apparatus.[47] Myers's own words provide the best description of this seminal event: "That there was any connection between the normal physiological reaction of a pilot and his lack of ability to do Blind Flying had never occurred to either of us until Major Ocker's experience with, and his absolute belief in, the action of the bank and turn indicator crashed head on into the author's knowledge of induced vertigo and the physiological reactions involved in the special senses concerned. Out of the wreck emerged several things of vital importance to aviation."[48]

The revelations of the so-called Ocker Vertigo Stopper Box transformed the nature of manned flight. Ocker and Myers made it clear that pilots could not trust their own sensations or instinctive responses: the human vestibular system led naturally toward vertigo and disorientation during blind flight. Instead, pilots needed special instruments to represent the environment for them. They also required extensive training to learn how to interpret flight instruments and use mechanical indications of turn, pitch, speed, and direction to create a mental picture of their orientation to the ground.[49] Total reliance on instruments, combined with realization of their own inherent inability to fly blind, allowed pilots to fly with machine-based precision and avoid succumbing to insidious vestibular illusions. Historians have explored "how machines can represent, communicate, and even create truths about the natural world."[50] Flight instruments played this role, providing an interpretation of the environment based on objective measurement. In this new approach to flight, pilots had to substitute their own physiological perceptions for the "truths" created by their instruments. This marked a new approach to flight that elevated the status of engineers and reduced the importance of a pilot's innate judgment. One way to understand the impact of these events is to employ the concept of paradigm change.

In his 1962 book *The Structure of Scientific Revolutions*, Thomas Kuhn described the nature of change in scientific knowledge and practice. To Kuhn, the traditional pattern of thought among a relevant community of practitioners represented a paradigm, and the normal scientific pursuits of the community occurred within this paradigm.[51] The paradigm was disrupted when anomalies appeared that subverted "the existing tradition of scientific practice."[52] Anomalies catalyzed a reconceptualization or reinterpretation of the natural world and the emergence of a new paradigm. Kuhn argued that the new paradigm surfaced only if it could solve a problem unanswerable in the old paradigm and if the transition to the new paradigm marked a scientific revolution.[53] Although Kuhn applied this model to the history of science, it offers insight into how change occurred in aviation's community of practitioners.

In the traditional paradigm of flight, a pilot's skills and seat-of-the-pants flying instinct helped prevent spatial disorientation. Although weather and darkness created nettlesome problems for navigation, competent pilots could employ their keen sense of balance to manage their way through blind-flight conditions, at least for short periods. Instruments such as inclinometers and magnetic compasses aided pilot judgment but could not be relied upon since they tended to go "haywire" in clouds or fog.[54] Weather-related losses of perfectly good aircraft (and aircrew), plus a pilot's inability to remain oriented in a hooded cockpit or a rotating Bárány chair, challenged the validity of normal practice. These inexplicable problems posed an anomaly, if not a crisis, for the dominant paradigm. When Ocker and Myers discovered that *pilots* were the core liability in blind flight, they ushered in a new paradigm that revolutionized flight.

Kuhn argued that a new paradigm won't emerge unless it solved a significant, acknowledged problem.[55] The realization that pilots were physiologically incapable of blind flight explained why airmen courted disaster once they lost sight of the horizon. It also established a more useful paradigm of flight: by trusting instruments instead of instinct, pilots could operate in blind-flight conditions for extended periods and thus expand the potential of military and civilian aviation. A major tenet of Kuhn's model of paradigmatic change and scientific revolution, moreover, suggested that a new paradigm would generate resistance from some members of the community of practitioners. And so it was in aviation. Although they had sired a revolutionary change in how humans fly, Myers noted that he and his colleague "were both promptly labeled as being enthusiastically crazy."[56]

Despite the verification of their results in airborne tests, Ocker and Myers faced a barrage of criticism.[57] Many pilots remained unconvinced that their own innate sense of balance rendered them incapable of blind flight, and they were unwilling to grant instrument technology authority over their own judgment. Indeed, a tradition of resistance against instruments was established not long after the first instruments appeared. The Wright Brothers, for example, installed basic instruments by 1914 but had trouble convincing their students to use them. The Wright's students were stung by the ridicule of other pilots, who considered instruments an unnecessary crutch.[58] Lieutenant Albert Hegenberger suggested in 1923 that aversion to instruments stemmed from minimal instrument use during flight training; since pilots did not require instruments when they learned how to fly in clear weather, they failed to appreciate their usefulness in foul weather.[59] Lieutenant Elmer Rogers, the Army Air Corps flight instructor who cautioned his students with the story of Lieutenant X's disastrous fog-flying mishap, noted that pilots resisted reliance on instruments to avoid the stigma of lacking in natural ability.

The commandant of the School of Aviation Medicine observed in 1945 that resistance to instrument-based flight persisted until the late 1920s because pilots preferred their own "God-given sense" and opposed the notion of becoming "mechanical" flyers.[60] The new doctrine of instrument-based flight demanded just that: pilots had to accept that the engineer's gyroscopes, needles, and dials were superior to even the most refined sense of human balance and judgment. Although many pilots were slow to embrace radical change, "mechanical" flying took great strides shortly after the revolutionary demonstration of the Ocker Vertigo Stopper Box.

The Technology of Instrument-Based Flight

Operationalizing the new paradigm of instrument-based flight required significant improvements in the science and technology of flight instruments. Safe flight in blind conditions required sensitive gyroscopes, reliable compasses, radio guidance systems, and precise altimeters. All of this, of course, required money and expertise. In the same month that Ocker and Myers made their breakthrough discovery, American industrialist and aircraft enthusiast Daniel Guggenheim turned his attention to expanding aviation's commercial capabilities. Poor weather reduced the utility of airmail aircraft, passenger aircraft, and military aircraft alike, and Guggenheim financed a civilian program in 1926 to tackle this issue. Endowed with $2.5 million, a staggering sum in the Roaring Twenties, the Guggenheim Fund for the Promotion of Aeronautics researched ways to dissipate fog or find a means to fly in it with safety and precision.[61] The fund enlisted guidance from the elite of American science and aviation. Its ten trustees included Nobel laureates Albert Michelson and Robert Millikan, plus Charles Lindbergh and Orville Wright, and they served until the fund disbanded on 1 February 1930.[62] Although the fund operated for just four years, Guggenheim's investment paid off in spectacular manner.[63]

On 24 September 1929, two Army Air Corps lieutenants made history on a flight sponsored by the Guggenheim Fund. James H. Doolittle occupied the front seat of a two-seat training plane, and Benjamin S. Kelsey sat in the rear. They taxied out to the runway, throttled up the engine, took off, and returned for a safe landing fifteen minutes later. Doolittle handled the controls the entire flight, but his cockpit was covered by a hood that eliminated all visual reference to the outside world. Kelsey could see outside but never touched the controls. Relying solely on his instruments, Doolittle had accomplished the first blind takeoff, flight, and landing. One observer described this feat as "the greatest advance in aeronautics since the Wrights first flew at Kitty Hawk."[64] Just as the Wright Brothers established a unique means of controlling a flying machine, the scientists, engineers, and aviators of the Guggenheim Fund's "Full Flight Laboratory" created another form of controlling an aircraft in flight. This new form severed the pilot's direct interface with the environment and interposed an artificial rendering of his surroundings.

Doolittle's blind landing required the use of several types of instruments invented or modified by the Guggenheim Fund. An instrument depicting the aircraft's relationship with the Earth's horizon enabled the pilot to ignore his vestibular sense and perceive his position in space. This so-called artificial horizon, a gyroscopically driven device fashioned by the Sperry Company, allowed small adjustments in the pitch and roll of the aircraft.[65] Another key instrument, the directional compass, also relied on its own gyroscope. Far more precise than the normal compass or even the Earth inductor compass, the directional compass allowed the pilot to maintain an accurate heading. Doolittle used this instrument in concert with another cockpit device that consisted of two parallel, vibrating lines. This device sensed the "localizer" beam emitted by a special radio beacon near the runway. The localizer beam provided an electromagnetic pathway to the runway. When flying in the center of the beam, both lines vibrated equally; if to the left or right, the left or right line would vibrate more than its partner.[66] Doolittle used the compass to maintain a correct heading parallel with the runway, and the localizer receiver provided a means to stay lined up with the runway. These instruments allowed Doolittle to point directly down the runway even though he could not see it, but he required additional instrumentation to determine his distance from the runway.

The Guggenheim researchers positioned the radio beacon a known distance from the runway. When the aircraft flew over the beacon the vibrating lines on the localizer indicator momentarily stopped vibrating. Upon receiving this indication, Doolittle reduced the throttle to initiate a controlled, 600-feet-per-minute descent.[67] This maneuver required a reliable altimeter to avoid slamming into the ground. The Guggenheim Fund used a precision altimeter designed by the Kollsman Instrument Company, the same company that later designed the space sextant and other precision instruments for the Apollo spacecraft.[68] Regular barometric altimeters could deviate by 100 feet, but the Kollsman altimeter was accurate to within 10 feet. At 50 feet above the ground, Doolittle eased the throttle back a bit more, made a slight increase in pitch to decrease his rate of descent, and glided along the localizer beam until he contacted the runway in a normal landing attitude. All of this required considerable concentration on the part of the pilot, and future developments in instrument flight made it easier for pilots to perceive and apply the information presented on their cockpit instruments.

Doolittle was a fitting choice for the first pilot to conduct an instrument-only flight. He not only possessed a Ph.D. in aeronautical engineering but served an active role in the Air Service's development of navigation techniques and all-weather flight in the early 1920s.[69] On a flight across Wyoming in 1924, for example, he encountered firsthand the problems of flying near thunderstorms. Sitting in an open cockpit, he was struck by hailstones that began to form a large snow cone on his helmet. Pummeled and freezing, and "fearing for the propeller and wings," he reversed direction away from the storm.[70] A few days later he experienced the danger of losing one's way in poor weather: "We decided that Seattle should be directly under us. We throttled the motor and had to glide through the clouds. Shortly before going through [the bottom of the clouds], a tree-capped peak brushed by directly under us. We zoomed [climbed] and then started feeling our way through again."[71] Apparently, his ability to use basic instruments prevented him from suffering the same disorienting effects as Lieutenant X. Doolittle's academic training and operational experience proved especially valuable, and his central role in the Guggenheim Fund symbolized the close connection between blind-flight research and the Army Air Corps.

The Guggenheim Fund's integration with the Air Corps paid dividends for both organizations. In his handwritten notes in the margins of the fund's official history, Lieutenant Albert Hegenberger pointed out the fund's heavy reliance on prior Air Corps expertise. When the fund declared that its Full Flight Laboratory was the first to give "competent advice as to the real needs of aeronautics in fog flying," Hegenberger penciled in "Bunk." He added, "The Materiel Division [at Wright Field, Ohio] has been broadcasting competent advice for years and gave the Fund *plenty*."[72] The Guggenheim Fund, apparently, did not emerge from a vacuum of knowledge. Instead, it built upon the experience and basic instrument technology acquired by the Air Corps.[73] When the fund terminated its work in 1930, it sent its airplane, radio beacon equipment, and other research material to the Air Corps' Materiel Division at Wright Field, Ohio.[74] Assistant Secretary of War F. Trubee Davison, a former trustee of the fund, declared in 1930 that "we have a moral obligation to carry on the [fund's] work," and he identified the Air Corps as the "best agency in the country to do it."[75]

Instrument flight and navigation became increasingly sophisticated in the decade before the Second World War. The notion that pilots must submit to a superior instrument-based method of flight gained greater acceptance as Army aviation improved blind-landing procedures, cockpit instrumentation, and training techniques. One of the Air Corps' first priorities was to transform the Guggenheim Fund's blind-landing experimentation into a reliable, standardized system usable for commercial and military operations.

The Air Corps sought to enhance the versatility of aircraft by developing a reliable instrument landing system that required minimal training for aircrew use and could be set up at any desired location. By 1931, the ground equipment included radio beacons and antennae portable enough for transportation in small vehicles or even motorcycle sidecars.[76] Once set up at specific distances from an airfield, this equipment transmitted information that enabled instrument-based guidance to the runway. The Air Corps also improved aircraft equipment. Cockpits were equipped with a new radio compass that pointed to the portable radio station and allowed pilots to orient themselves to the runway.[77] Captain Albert Hegenberger used this equipment to establish another first in aviation. On 9 May 1932, he climbed into a hooded cockpit, took off, flew around the local area, and landed without ever seeing the ground. This resembled Doolittle's landmark sortie of 1929, but Doolittle had a safety observer. Hegenberger flew alone. This hooded solo flight confirmed the reliability of the new system, and Hegenberger's official report declared that the average pilot could achieve proficiency "in a reasonable short time."[78]

A month after his blind solo flight, Hegenberger received a letter from Billy Mitchell. Mitchell, a staunch airpower enthusiast even after his 1925 court martial, sensed the importance of Hegenberger's accomplishment: "I notice you have been doing some very interesting instrument flying lately. I would like very much to hear about what your instruments were and what you actually did."[79] Apparently, Mitchell sensed that conquering blind flight created delightful opportunities for the application of the air weapon. In later correspondence,

Mitchell requested further information from Hegenberger on the specifics of instrument advances and observed that "we are making a good deal of headway in our endeavor to do something with aviation and national defense."[80] Hegenberger replied: "As far as we are concerned fog landing was solved in May 1932 [on the blind solo flight]. Its solution proved far simpler that [*sic*] we anticipated, both from the standpoint of the pilot's functions and in the necessary equipment."[81] The Air Corps shared Hegenberger's confidence.

In his annual report of 1933, the chief of the Air Corps Materiel Division declared that radio-guided instrument landings achieved a "practical stage" in 1932, and numerous Air Corps pilots validated its accuracy during winter weather conditions.[82] This set the stage for the endorsement of the Air Corps' "Blind-Landing System" by the Department of Commerce.[83] The Air Corps demonstrated its instrument landing technology to Department of Commerce representatives at Langley Field, Virginia, in October 1933. The impressed Commerce officials established the Air Corps' system as the industry standard, and the Air Corps began to train Commerce inspectors as well as a cadre of pilots from commercial aviation firms such as Transcontinental and Western Air (TWA).[84] The Air Corps also rushed to set up instrument landing systems to aid in a new mission: hauling the mail.

In response to a scandal involving the US Postmaster and civilian airline executives, Franklin D. Roosevelt assigned the Air Corps responsibility for the nation's airmail in the winter of 1934. The Army had performed this mission for a few months in 1918 but on a much smaller scale than required in 1934.[85] Despite substantial progress in instrument flight during the previous several years, the Air Corps' attempt to take over the airmail routes resulted in a series of embarrassments. In only seventy-eight days, the Air Corps suffered sixty-six crashes and twelve fatalities.[86] Three separate crashes on one day, 9 March 1934, resulted in four deaths.[87] Several mishaps resulted from a combination of inadequate instrumentation and insufficient training in blind flight procedures. Most Air Corps pilots still received little instrument training, and those who were competent in instrument procedures were unlikely to be familiar with the new instrument landing system. Yet the airmail fiasco bore some fruit: media attention loosened federal purse strings, the Air Corps gained funds and impetus to institutionalize instrument training, and pilots realized that their ability and reputation depended on mastering the new paradigm of instrument-based flight.[88]

Instrument Indoctrination

The new paradigm of instrument-based flight transformed the nature of pilot training. Gone were the days of trying to improve pilots' blind-flight capability by spinning them around in the Bárány chair or whirling them about in all three axes in the Ruggles Orientator. Recognition of the human incapacity to fly blind rendered such practices of the early 1920s obsolete. The experiments of Ocker and Myers, the pioneering work of the Guggenheim Fund, and the achievements of Doolittle, Hegenberger, and others at the Materiel Division in the late 1920s and early 1930s made it clear that the only way for aviators to overcome the deadly effects of vertigo was to abandon reliance on their own senses and adopt the disciplined procedures of instrument flight.[89]

The airmail mission hastened development of a formal school of instruction in instrument procedures at Wright Field during the spring of 1934. The chief of the Materiel Division hailed this as the "outstanding accomplishment of the year."[90] The program began with twenty pilots who learned how to conduct blind landings in small trainer aircraft as well as the larger Martin B-10 bomber. It was a small but measurable start. Stung by the airmail failures, the chief of the Air Corps decreed in 1935 that *all* Air Corps pilots must complete a formal course of instrument-flying training. This included at least twelve hours of flight instruction in a hooded cockpit as well as periodic tests to "determine their fitness for instrument flying."[91] Only after pilots achieved basic instrument proficiency could they attempt to master more complex maneuvers such as blind landing.

Basic proficiency in instrument use required pilots to master a simple approach to flight, the "1-2-3 method,"[92] which required a minimal complement of cockpit instruments: a turn needle that swung left or right when the aircraft changed its heading, a bank indicator that displayed how much the wings tilted to the left or right, an airspeed indicator, and a compass. Pilots learned to maneuver without reference to the ground by focusing on these instruments and applying a step-by-step, mechanical method. A pilot had to (1) step on the rudder pedals to move the turn needle left or right, which moved the nose left or right; (2) position the control stick to either side to make the bank indicator move, which tilted the wings left or right; and then (3) pull back or push forward on the stick to make the airspeed decrease or increase, which pitched the nose up or down. This method enabled pilots to maintain straight and level flight, fly precise turns, prevent spiral dives or spins, and perform controlled glides under blind-flight conditions. Had Lieutenant X been trained in this technique, he may have avoided his deadly spiral toward the ground.

As instrument technology advanced through the 1930s, the cockpit became festooned with a variety of gadgets: altimeter, artificial horizon, bank indicator, compass, directional gyro, fuel pressure gauge, localizer receiver, marker beacon indicator, oil pressure gauge, radio equipment, tachometer, turn needle, and vertical velocity indicator, to name a few. These various instruments competed for space and attention, and the Materiel Division established a standard cockpit instrument arrangement to maximize utility for the pilot.[93] At the center of the instrument array was the gyroscopically driven artificial horizon. Doolittle's famous 1929 blind flight demonstrated the merits of this device, and it was pressed into wider service throughout the 1930s. By the Second World War, it was standard in most cockpits and became the central focus for pilots during blind flight.

The artificial horizon approximated the location of the actual horizon. To aid the pilot's situational awareness, a small airplane silhouette was superimposed in front of the artificial horizon. This enabled the pilot to perceive his or her position relative to the ground. Using this device, the pilot manipulated the controls to alter the relationship between a tiny, artificial airplane and an artificial horizon. This reductionist approach was aided by the needles and numbers displayed on surrounding gauges and dials. Pilots learned to process this information and translate it into appropriate control inputs that enabled safe passage through weather or darkness. This redefined the pilot's role: he no longer flew the *aircraft*; he flew the *instruments*.

The purposeful, user-friendly cockpit arrangement of increasingly sophisticated flight instruments improved a pilot's efficiency and reliability. By World War Two, pilots were no longer taught the old 1-2-3 method, which relied on just a few basic instruments and demanded intense concentration. Instead, they were taught the "full panel" system in flight school. This system encouraged pilots to use all of their instruments and develop the ability to fly in limited-visibility conditions for extended periods with relative ease. The goal was to help pilots override their instinctive physical sensations by cultivating an "instrument consciousness."[94] To this end, the full-panel system emphasized the science of spatial disorientation. The Army Air Forces 1943 basic manual for instrument flight, for example, employed cartoons and diagrams to explain the physiology of the inner ear and caution airmen about the hazards of vestibular illusions.[95] Equipped with this scientific knowledge, airmen could reason away the insidious sensations that appeared during blind flight.

An instrument-conscious pilot was a predictable, disciplined pilot. He or she could be relied on to interpret the instruments correctly and administer appropriate control inputs, even when fatigued, stressed, or spatially disoriented. Mastering complex procedures such as the instrument landing system required considerable effort, and pilots were commanded to practice instrument-based maneuvers "innumerable times" to gain complete confidence.[96] The goal was to develop a habitual, automatic approach to flight.[97] This referenced a theme in instrument flight: repetition becomes intuition. The more a pilot used his instruments, the greater trust he placed in them. A competent instrument pilot placed little trust in his own physical sensations of flight and relied instead on the analog interpretation of reality provided by a suite of finely calibrated, pressure-sensitive, gyroscopically powered instruments.[98]

When experiencing vertigo or vestibular illusions, the instrument-conscious pilot should, according to a World War Two instrument training guide, "say to yourself—'there I go again,' and if possible laugh at yourself."[99] While pilots may have been laughing off disorienting illusions, they were also surrendering more and more of their own judgment and authority to external control. Onboard instruments directed the thoughts and actions of pilots, and so did mechanisms and agents exogenous to the aircraft. Ground-based devices such as radio beacons told the aircraft where to point, and localizer beams that emanated from the runway guided aircraft to a safe landing. Although pilots retained ultimate responsibility for the safety of their aircraft, advances in the science and technology of radio communication and radar guidance further eroded their autonomy.[100] This occurred in the form of a wartime innovation produced by the Radiation Laboratory at the Massachusetts Institute of Technology.

In 1941 an MIT physicist conceived a new method for landing aircraft in low-visibility conditions. Dr. Luis Alvarez, a future Nobel laureate, surmised that radar's ability to track an aircraft's distance, elevation, and azimuth (side-to-side angle) could be used to guide aircraft to a runway at night or in poor weather. Radar operators on the ground radioed directions to an aircraft and, in essence, talked the pilot down to the runway. The Radiation Lab called this procedure a *ground-controlled approach* (GCA), a term and practice still used today.[101] Unlike the complicated instrument landing system developed by the Air Corps in the early 1930s, GCA placed little burden on the pilot. "Instead of putting the information on [an instrument] in front of the pilot," Alvarez explained, "it is told to him verbally over his regular communications radio."[102] The pilot obeyed simple instructions from a ground controller, such as "turn right," "turn left," "begin descent," "on course," "on glideslope," etc.[103] One Army general noted in early 1943 that GCA could be used with great success by pilots completely unfamiliar with the system, and another officer predicted its usefulness to aircrew landing from long sorties in unfamiliar locations.[104] By the end of World War Two, the American military operated ninety-three GCA units, more than half of them overseas.[105] When bad weather rolled in over the landing field and returning aircraft were out of fuel and out of options, the system saved numerous airmen from the unhappy alternative of bailing out.

Perhaps the most telling perspective on GCA came from a 1947 report from the Air Materiel Command. This report praised GCA since "the pilot would have his thinking done for him by someone on the ground."[106] Although the pilot still played an active role during a GCA approach as an actuator of throttle, stick, and rudder, his situational awareness was also supplied to him by an external agent. This fit with the new paradigm of instrument flight established by William Ocker and David Myers two decades earlier. During blind-flight conditions, pilots could not maintain their orientation using their own senses and judgment. Instead, they had to submit to the new authority of a mechanical interpretation of reality and, in the case of GCA, a voice over the radio.

Despite developments such as GCA, pilots possessed ultimate responsibility over their aircraft. Instrument-based flight still enabled the pilot to retain his status as the primary agent of control. Historian Erik Conway notes that even after technological developments enabled completely automatic, hands-off "blind landings," many pilots preferred other instrument-based landing techniques that still required pilot judgment and physical inputs. Such preferences illustrate how pilot culture shaped aviation technology. Just because technology existed that could automatically land the aircraft did not mean that pilots would choose to use it. As Conway argues, most pilots were unwilling to risk the most critical phase of flight—the landing—to a black box.[107] Nevertheless, the point here is that exploiting the potential of manned aviation required fundamental changes in how pilots flew their aircraft. Flight became the process of manipulating cockpit controls to effect changes on cockpit instruments or comply with external commands. Only then could a pilot perform as a reliable component of the technological system.

Conclusion

Blind flight provoked a clash between human physiology and military and commercial necessity. Pilots were unable to remain oriented in fog, clouds, and darkness, and this placed severe constraints on the use of aircraft. Although medical science had described the basic mechanisms of vestibular physiology by World War One, until 1926 no one suspected that a normal sense of balance prevented pilots from operating in blind-flight conditions. This discovery was precipitated by the development of a gyroscopically powered flight instrument and the insightful observation of a flight surgeon. The realization that pilots could remain oriented only if they ignored their own senses and trusted special instruments changed the nature of flight. Once airmen grasped the necessity of instruments, the development of blind-flight capabilities proceeded apace, and flying went from a multisensory operation to a multiinstrument operation. In the new paradigm of instrument-based flight, pilots submitted to the authority of an artificial representation of the environment. This signaled another shift away from the tradition of pilot supremacy and toward greater reliance on engineers and an increasingly complex technological system.

As with the problems of high altitude and G-forces, solving the problems of blind flight involved a multifaceted interplay between science, technology, and military necessity. Overcoming the reverse salients imposed by environmental extremes and blind-flight conditions required new scientific knowledge of human function, yet this knowledge was enabled by specific technological achievements. The altitude chamber revealed much about respiratory physiology, and Elmer Sperry's gyroscopic turn indicator helped expose the relationship between spatial disorientation and vestibular physiology. Furthermore, the imperative to advance aviation capabilities fostered new technological and scientific developments. The need for bombers to obtain high altitudes helped spur the creation of pressurized aircraft and promoted research in the effects of low atmospheric pressure. Similarly, the military necessity to function in all types of weather promoted development of radio beacons, localizer beams, and radar-guided approaches.

The development of instrument flight also facilitated improvements in the automatic control of aircraft. By 1937 the Air Corps integrated its instrument landing system with an aircraft autopilot and achieved another first: an aircraft that performed a fully automated approach and landing in adverse weather conditions without any input from a pilot.[108] Although this was a manned flight, the pilot was little more than a passenger. Major General Oscar Westover, the chief of the Air Corps, hailed this event as "the last great step in eliminating flight hazard incident to zero visibility."[109]

Chapter Four

The Changing Role of the Human Component

The models of technological change discussed in the preceding chapters reveal little about actual human–machine interaction. Thomas Hughes's concepts of technological systems and reverse salients inform the effects of human physiology on aviation development, and Edward Constant's notion of presumptive limits provides another perspective on revolutionary change in aviation technology. Yet these models overlook the substantive interplay between flyers and their machines. During the interwar years and through the Second World War the air force modified the role of the human component, and these efforts anticipated the post–World War Two concept of cybernetics.[1]

Essentially, while trying to improve the air weapon, the predecessors of modern aviation presaged cybernetics, or at least an early form of it, before it received formal conceptualization in 1948. Cybernetics thus provides a useful model to understand the changing interaction between airman and aircraft in aviation evolution. This interaction was also informed by various phenomena such as human factors engineering, new perspectives on human error, and training methods that standardized the function of the human component. As aviation evolved, moreover, new devices appeared that proved superior to the human component and usurped many of the airman's traditional responsibilities. This trajectory toward greater automation, along with changes in the integration and function of the human component, helped redefine the pilot's role in flight.

Cybernetics

Norbert Wiener, a professor of mathematics who worked on automation technology in World War Two, coined in 1947 the term *cybernetics* (from the ancient Greek word for navigation by a steersman or, more generally, governance) to describe "the entire field of control and communication theory."[2] Basically, systems that sense and react to external inputs are cybernetic systems. A key aspect of this process is information feedback: cybernetic systems convert information to action, this action generates information used for future action, and the system continues to self-regulate as information is fed back into it. Simple examples are a heater hooked to a thermostat, or a speed governor on an engine.[3] In these systems the goal is a state of equilibrium or homeostasis that serves a designed purpose such as maintaining a precise temperature or specific operating parameters. Other cybernetic or information feedback systems are more complex; indeed, Wiener noted that adequate computing power and proper sensing mechanisms could provide a cybernetic system with "almost any degree of elaborateness of performance."[4] One example of a complex cybernetic system comes from Wiener's own work in the early years of the Second World War.

Using ground-based guns to shoot down distant, fast-moving aircraft posed an immense challenge, and Wiener worked from 1940 through 1942 on improving the accuracy of antiaircraft artillery. Accuracy depended on the ability to predict the future position of enemy aircraft and direct fire to that location. With the aid of an electrical engineer, Wiener devised a circuit board that computed an aircraft's flight path and predicted its future position. The gunner tracked the aircraft with an optical sight that furnished information to Wiener's predicting computer, which calculated the flight path based on these data.[5] This enabled a feedback process: the predictor estimated the flight path, and as new information came in, the predictor updated its estimate based on the observed position.[6] Wiener surmised that the longer the gunner could track the target, the more accurate a prediction could be made of its future position, and the gunner could then point the gun at the target's projected location and fire.

Essential to Wiener's analysis was his observation that the pilot "behaves like a servomechanism" effecting predictable control inputs based on physical and visual sensory data.[7] Similarly, historian David Mindell characterized the gunner as a "manual servomechanism," or a machine within the larger machine.[8] The accuracy of Wiener's fire control system relied on the gunner's ability to track the aircraft with an optical sight: the gunner translated information from the computer into physical action by slewing the muzzle to the predicted position. In addition to illustrating the importance of feedback-based control, the antiaircraft device thus demonstrated how human components could function within a complex cybernetic system.

Overall, Wiener's wartime work in antiaircraft artillery showed how system performance could depend on human and machine components and their interplay with each other. This suggested that the gunner or the pilot could be replaced by other servomechanisms, either human or machine. Accordingly, after Wiener introduced cybernetics as a means to study communication and control, he contended that, within the context of a cybernetic system, the essential functions of human and machine components could be understood in the same mechanical terms.[9] This implied that humans could be relegated to the status of an organic component within a complex cybernetic system. Optimal function, however, required smooth integration of the human component with the rest of the system.

The Cybernetic Nature of Flight

The basic cybernetic principles of control, communication, feedback, and self-regulation, combined with Wiener's treatment of the human component as another mechanism within a larger system, provide insight into the evolving airman–aircraft complex in the early decades of military aviation.[10] As Wiener himself noted, "Although the term *cybernetics* does not date further back than the summer of 1947, we shall find it convenient to use in referring to earlier epochs of the development of the field."[11] Although the principles of cybernetics received a thorough conceptualization after the war, they were in full practice decades earlier.

Even before a mechanical form of control and stability took root in aviation, Elmer Sperry demonstrated the superior qualities of gyroscopic control at sea. The Sperry Gyroscope company developed a gyrocompass, or Gyro Pilot, that controlled ship steering far better than the traditional "human steersman."[12] Control of the ship was no longer effected directly through the sailor's hand. Instead, a ship's pilot managed and adjusted the gyroscopic machine to ensure it continued to do the job better than its human predecessor. Sperry's application of gyroscopic control expanded beyond ships and produced increasingly sophisticated autopilots and control instruments in aircraft. The institutional embrace of these devices anticipated the concepts that Wiener would later label as cybernetics. In this vein, consider how information was sensed and acted upon to achieve desired parameters in flight.

Before the development of reliable instruments and autopilots, the pilot served as the primary means of sensing the environment and controlling the system. In the first years of flight airmen judged their speed by sight, sound, and feel. A pilot could sense a dangerous loss of airspeed by observing his progress over the ground, listening to the wind rush through the wires, or feeling a characteristic pre-stall shudder in the wings. These sensations were supplemented with crude airspeed instruments by the First World War, but imprecise perception of the environment still invited disaster. Second Lieutenant J. F. Woodruff's close brush with death while flying his biplane on 12 January 1918 illustrates the point.

While practicing simple turning maneuvers 400 feet above the ground, Lieutenant Woodruff failed to sense a drastic loss of airspeed, and his aircraft stalled and promptly spiraled into the terrain. The bruised but lucky lieutenant walked away from the accident only because he crashed in thick vegetation in a swampy ravine.[13] His mishap showcased the perils of an inefficient information feedback system. The organic component's inferior ability to sense changes in velocity resulted in the system's failure to regulate its speed through the air. A mechanical airspeed indicator might have provided more accurate information in time for him to prevent the crash. The introduction of reliable airspeed indicators enabled pilots to make timely decisions, and other improvements in instrument technology augmented the pilot's ability to sense and synthesize information. Recall how gyroscopic-based instruments replaced inadequate human senses and allowed flight through inclement weather. Overall, improvements in the ability to sense and apply information about the environment helped increase the "elaborateness of performance" of the cybernetic system.

Improving the performance of the airman–aircraft cybernetic system required exploiting the strengths and mitigating the weaknesses of each component in the system. Humans excelled at some tasks, machines excelled at others. Machine-controlled feedback systems such as automatic pilots performed far better than pilots at maintaining steady parameters.[14] Gyroscopically stabilized automatic pilots flew the aircraft in a stable, reliable manner, and they compensated for the pilot's tendency to become disoriented or fatigued during flight through inclement weather. Automatic control was also more precise. Basic flight maneuvers such as turns, climbs, and straight-and-level cruise could be performed by the human pilot or the automatic pilot, but the autopilot operated with greater speed and accuracy than its human counterpart.[15] This superior form of control proved particularly useful in stabilizing the aircraft's flight path during precision bombing runs. The pilot's manual for the B-17 Flying Fortress during World War Two noted that the autopilot "detects flight deviations the instant they occur, and just as instantaneously operates the controls to correct the deviations."[16] This was a classic example of how an information feedback system maintained steady parameters, and it provided a direct analogue to the systems cited by Wiener in his 1948 treatise on cybernetics. Despite their accuracy and stability, however, self-regulating control devices such as the gyro-stabilized autopilot suffered a significant weakness: these components could not maintain control of the system during dynamic, unpredictable maneuvers. Consider the practice of V-1 tipping.

The Vergeltungswaffe-1 (Vengeance Weapon 1, or V-1) employed by Germany in 1944 and 1945 was an unmanned cybernetic system with a deadly payload. It looked like a small airplane and proceeded on an unwavering flight path toward its target with the aid of an internal gyroscope that sensed minor deviations and directed control inputs to maintain a steady, stable course. After intercepting a V-1, Allied pilots had two options: shoot it or "tip" it. The sport of V-1 tipping involved sidling up next to the device, placing the fighter's wing under the V-1's wing, and then tipping it over. According to Lieutenant General James H. Doolittle, the commander of the Eighth Air Force during the V-1 attacks, this maneuver "upset the gyro, and the V-1 would spin in before arriving at England."[17] V-1 tipping illustrated that erratic maneuvers exceeded the capabilities of mechanical control devices of the era. It demonstrated that pilots were necessary for flight that exceeded certain parameters. One surmises that the integration of a pilot into the V-1's control loop by either direct presence or remote control would have ended the practice of tipping.

The human component improved system function by providing a means of control during unpredictable flight operations. Autopilots failed in strong turbulence and were useless in an aerial dogfight. Pilots supplied a general awareness of the environment that exceeded machine capabilities. A pilot was required to steer around thunderstorms, outmaneuver enemy threats, maintain position in a formation of aircraft, or compensate for various malfunctions. On many occasions the airman's ability to improvise and make judgments was critical to survival.[18] Captain David Wright's experience as a medical observer aboard a B-17 sortie over Germany provided an example of such flexibility. On a normal mission, the B-17 would have relied on a mechanical autopilot for a precise bomb run and to reduce pilot fatigue, but on this mission the bomber was attacked before reaching its target. German fighters riddled the Flying Fortress with bullets, knocking out control systems and blowing large holes through the wings, two of the four propellers, the fuselage, and the nose. Wright witnessed the ten-man crew, nearly all of them wounded to some degree, struggle to keep the aircraft flying as it dropped out of formation and limped home across the English Channel.[19] In this case, the human components were able to react to a dynamic situation and provide control inputs that stabilized the operation of the system.

The hallmark of a cybernetic system is the ability to receive, process, and apply information to effect control.[20] Both human and machine components played important roles in complex cybernetic systems such as the B-17 and other aircraft. Aircrews perceived and responded to dynamic conditions, and gyroscopic-based mechanisms offered reliable, precise control during stable, routine operations. Leveraging the different strengths of the organic and inorganic sensing and control mechanisms improved overall function, and aircraft performance depended on the efficient integration of the human and machine components. Cybernetics, therefore, underlies the evolving interplay between airman and aircraft and informs the story of how the early air force shaped the roles and interaction of human and machine components in the evolving technological system of military aviation.

This story comes in several parts. It begins with the notion of *human factors engineering,* or the air force's effort to integrate the human machine into an efficient man–machine complex. Human factors engineering expanded on the work done by flight surgeons and included improvements in safety and efficiency as well as new perspectives on pilot error and designer error. Equally important were the air force's efforts to standardize the human component and to improve the machine components. Compensating for the inherent variations among human operators required various techniques to ensure that airmen performed in a predictable, proficient manner. However, improvements in the machine components would soon shift the balance toward greater machine control, altering the airman's role in the cybernetic system of flight.

Integrating the Human Component
Human Factors Engineering

An early account of overcoming a problematic airman–aircraft interface involved Lieutenant Benjamin Foulois in 1910. Foulois was one of the initial pilots of *Signal Corps No. 1,* the U.S. Army's first military airplane. The Wright Brothers built *No. 1* and provided flying lessons to Foulois and other officers. During a solo flight on 12 March 1910, Lieutenant Foulois hit some gusty winds and was thrown toward the top of the aircraft: "The two truss wires in front of the pilot's seat saved me from being thrown completely out of the airplane. . . . Thereafter, I used a four-foot trunk strap with which I lashed myself to the pilot's seat."[21] Foulois's impromptu addition of a safety belt marked one of the earliest, simplest efforts to enhance system performance by improving the pilot's integration with the machine. Similar and increasingly sophisticated efforts occurred throughout the interwar years, but it was not until the end of World War Two they were labeled as *human factors engineering*.[22]

To some extent, the previous chapters can be viewed in terms of human factors engineering since flight surgeons and aviation engineers helped to design machines that fit the human component. Human factors engineering also involved the physical interaction between airman and aircraft in areas other than life-support systems and flight instruments. To improve safety and efficiency, it incorporated into aircraft design scientific data about human behavior, cognitive ability, and physical limitations. This informed the development of aircraft as cybernetic systems since humans had to be properly integrated into the machine in order to sense the flight environment and exert control. Thus, although the term may not have appeared until the mid-1940s, the practice appeared in various forms during the previous decades and dramatically shaped the pilot-aircraft complex.[23]

During the early 1920s, engineers assimilated human factors into aircraft design by translating subjective opinions about flying qualities, or how an aircraft "feels," into adjustments that affected aerodynamic stability. The symbiosis between pilot and aircraft involved striking a balance between the aerodynamic properties of the aircraft and the pilot's perception of how it handled during flight.[24] The Army's 1921 *Handbook of Instructions for Airplane Designers* addressed this relationship between pilot input and control response when it advised engineers to ensure that flight control movement did not require undue pilot exertion.[25] By locating official concern for human physical limitations outside of the tradition of aviation medicine, the *Handbook* provided early guidance relevant to human factors engineering.

Designing machine components to improve the efficiency, safety, and comfort of the human operator involved many aspects of flight. In 1921 the Army Air Service's Engineering Division recommended that the Martin MB-2 bomber be equipped with a windshield to protect the pilot from rain, hail, and windblast.[26] The absence of this basic device from the original design suggests that engineers tended to overlook the human component. According to "Ideal Specifications for Military Airplanes," an engineering guideline commissioned by the chief of the Air Service in 1923, Army aircraft should have adjustable seats to accommodate parachutes, and the seats should be far enough behind the propellers to avoid injury if the aircraft noses over during a crash. Aircraft seats should also be located so that only in the unhappy event of "a straight nose dive into the ground" would the engine and airman crush together. Additionally, all seats and even the crude instrument panels of the era should be designed for ease of use and maximum aircrew comfort.[27] These modifications were consistent with one flight surgeon's 1924 observation that "the engine and controls of the airplane must be made a part of the aviator's own body."[28] This sentiment portended Norbert Wiener's observation decades later that, when "coupled into the fire-control system," the human component must act "as an essential part of it."[29]

Efforts to improve the interaction between airman and aircraft involved more than just the pilot. Air Corps studies in 1932, for example, evaluated bombsight prototypes for their ease of use. Bombardiers flew operational tests with different types of bombsights and answered such questions as: "Are the operating controls of the sight located conveniently? Can the sight be operated satisfactorily with gloves? Is the height of the eyepiece above the floor satisfactory? Is the time interval between the warning lights and the dropping signal of satisfactory length?"[30] Overall, bombing experiments in the early 1930s illustrated that optimizing the interface between the bombardier and aircraft equipment improved bombing accuracy. Adjustments in the bombsight telescope's field of vision aided precision, and equipment was also developed to test automatic electrical release versus manual release.[31]

Bombardiers also found that they could hit their targets more often if an automatic pilot instead of a human pilot was used to make delicate course corrections during the bomb run.[32] This resulted in a synchronous interface between bombardier, bombsight, and autopilot that usurped control from the pilots during the most critical moments of the mission, leaving the pilots with a new role: sit and observe.[33] This once again anticipated Norbert Wiener's notion that cybernetic systems could perform elaborately complex functions.[34]

By the mid-1930s, the need for human-friendly design had become even more apparent. The 1934 version of the *Handbook of Instructions for Airplane Designers* directed engineers to arrange cockpit controls and instruments for "maximum comfort and convenience."[35] It also increased the impetus for engineers to standardize cockpit controls and construct seats and restraining devices to improve aircrew safety.[36] Safety improvements also required engineers to consider the physics of airplane crashes and design means to improve the chances of aircrew survival. The abundance of avoidable casualties in the

Second World War suggested that this aspect of human factors engineering was often overlooked or ignored. The N-3A gunsight installed in World War Two pursuit (or fighter) aircraft such as the P-39 and P-40 provided an example of dangerous design that gave little consideration to pilot safety. The gunsight protruded eleven inches from the instrument panel, a glass and metal cudgel ideally positioned to inflict massive head and facial injuries during an abrupt stop.[37] Similarly, the P-51's K-14 Gyro Gunsight, another electromechanical augmentation designed to help pilots hit their targets, posed a danger of hitting the pilot. It stuck out so far that it injured pilots even when their safety straps restrained the torso during a crash landing.[38] Such application and *mis*-application of human factors engineering during the interwar years and World War Two illustrated the need to incorporate all aspects of the human component into an efficient man–machine complex.

On the basis of experiences in the Second World War, the Army Air Forces established the Psychology Branch in the Aero Medical Laboratory at Wright Field in 1945. This new organization's charter was to facilitate the assimilation of human considerations into equipment design in order to enhance system performance. During the war, for example, bombing accuracy failed to meet expectations, as did gun control systems in fighters and bombers.[39] The Psychology Branch tried to figure out why. One of its studies determined that the complex gunsight controls in the B-29 Superfortress were haphazardly arranged and overly complicated; since most adjustments required use of the left hand, investigators surmised that the designer had been left-handed. Experiments with a more user-friendly device demonstrated significant improvements.[40] This sort of work fit the philosophy of Lieutenant Colonel Paul M. Fitts, an Army psychologist and the first chief of the Psychology Branch. In a foreshadowing of the cybernetic model, Dr. Fitts emphasized that aircraft performance depended on the airman's ability to perceive and manage information.[41] Equipment design, therefore, should be tailored to both psychological and cognitive capacities as well as physical characteristics.

An important but confounding aspect of the Psychology Branch's work and other efforts in human factors engineering involved the notion of variation. A pilot's physical and cognitive capabilities varied significantly not only from other pilots but over time as well. No two pilots were alike, and no single pilot was consistent. Just as bronze or steel cannons could burst when fired due to metallurgical impurities and variations, the variable human component could fail unpredictably during extreme operating conditions such as low barometric pressure, excessive acceleration forces, high temperature, and stress. As early as 1918 British investigators had observed that diminished oxygen pressure was met with "remarkable variation . . . among individuals who are apparently equally healthy."[42] Flight surgeon Harry G. Armstrong confirmed twenty years later that tests of altitude tolerance revealed significant variations among airmen.[43] Armstrong also noted that "only a shade of variation in human performance" could mean the difference between success and failure in military aviation.[44] With regard to the interaction between pilot and aircraft, however, there was one important variation that did not seem to make a difference: sex.

The Army Air Forces decided by 1942 to train women pilots to ferry its aircraft from the factory to military airfields. By the end of 1944, nearly 1,100 Women's Air Force Service Pilots, or WASPs, had flown over 600,000 hours in bombers, fighters, transports, and trainers.[45] On 7 December 1944 the chief of the Army Air Forces, General Henry "Hap" Arnold, addressed a formal gathering of WASPs. He confessed that, "Frankly, I didn't know in 1941 whether a slip of a young girl could fight the controls of a B-17 in the heavy weather they would naturally encounter in operational flying. . . . It is [now] on the record that women can fly as well as men."[46] The WASPs' director, Jacqueline Cochrane, reported that a woman—even a slip of a young girl—possessed adequate muscular strength to perform all pilot duties, and rates of flying fatigue among women compared favorably with those of their male counterparts.[47]

Although little variation seemed to exist between male and female regarding piloting ability, some questioned if the normal physiological variations due to the menstrual cycle would detract from safe, consistent performance. Dr. R. E. Whitehead, the chief of the Medical Section at the Bureau of Air Commerce, asked in 1934: "Did women pilots tend to be boyish females, homosexuals or a normal type and how were normal women affected by ovarian functions or menstrual change?" Whitehead suggested that a series of airplane crashes involving menstruating women implicated menses as a cause of poor pilotage.[48] He also advised that women should not fly within three days of the beginning or end of their period in order to minimize the risk of fainting.[49] Others disagreed with this view. The *Journal of Aviation Medicine* reported in 1941 that menses had no ill effect on piloting abilities so long as it was a "normal" cycle.[50] Jacqueline Cochrane's report on WASP performance seemed to settle the issue: investigations of 123 major accidents found no connection with the female cycle, and a medical study detected only a "slightly noticeable" change in concentration, coordination, or tenseness in less than one-fifth of its participants.[51]

The WASPs demonstrated that an efficient human–machine interface in aviation did not disqualify female "airmen." Yet it did require an appreciation of the varying physical and psychological features of airmen. Such variations in the human component imposed significant problems for engineers. The ideal airman existed only on paper—the real airman came in a wide range of physical and psychological forms. Indeed, airmen were troublesome, inconsistent components that defied standardization. An airman's inner workings resisted understanding or prediction, yet they were important in the input, processing, and output of information and action in the cybernetic system of flight.[52] Discovering how airmen managed to sense, assimilate, and apply information posed a major challenge for aviation engineers, and integrating the human component into the rest of the machine required careful design. Once the human was inserted into the technological system, errors of the pilot and errors in design became intertwined. The evolution of aviation is in large measure the story of how people identified and corrected these errors.

Pilot Error

Flight was unforgiving of inattention, ineptitude, or incaution on the part of the human component. Take, for instance, the misfortune of Second Lieutenant Keith Roscoe. On 11 July 1928, Lieutenant Roscoe flew in a formation of open-cockpit, single-seat Curtis P-1 biplanes. Flying in the lead position of the formation and cruising along at just 300 feet above the ground, Lieutenant Roscoe apparently felt the urge to sneeze. Unhooking his safety belt in order to reach his handkerchief, he sneezed while unrestrained, bumping the controls and causing a sudden pitch forward of the aircraft that flung him out of the cockpit. Roscoe's wingmen now beheld the curious sight of their formation leader no longer seated at the controls, but clinging to the outside of the aircraft as it dove toward the ground. They could only look on as Lieutenant Roscoe held on with one hand gripping the cowling under the small windshield and the other hand groping into the cockpit in a desperate attempt to grab the controls. Unable to reach the controls, he let go and tried to use his parachute, but it was too late.[53] Roscoe, as well as thousands of other flyers, paid the ultimate penalty for human error.

Between Pearl Harbor and V-J Day, the US Army Air Forces lost more than 14,000 airmen due to aircraft accidents in the continental United States alone.[54] Noncombat sorties killed, on average, about ten airmen per day. In April 1945 there were 732 major accidents, or about one for every hour of the day.[55] These numbers dwarfed the record of mishaps from the preceding decades, although the rate of accidents during World War One and the interwar years was even higher than in World War Two. In the Great War, 164 American aviators were killed in action, while accidents claimed 319, and 200 were classified as "missing in action."[56] From the beginning of the war through 1919, moreover, the Army recorded 1,250 crashes at various airfields in the United States.[57] From 1920 through 1936, 6,234 aircraft accidents claimed 544 lives.[58]

Nearly all of these accidents resulted from some form of human error. Many were due to simple carelessness, as in the case of Lieutenant Roscoe. Other accidents resulted from a lack of skill or discipline. Lieutenant Roland Birnn recorded such a case in his diary for 17 March 1942: "Twenty P-40's came through today. . . . One of the pilots, just out of flying school, rolled his P-40 just over the field and into the ground. Just doesn't pay to get hot."[59] In this case, the young pilot's foolhardy attempt to impress observers on the ground cost him his life. Some accidents resulted from errors on the part of maintenance personnel, engineers, or the manufacturer. In a personal account of his World War One flying experiences, Lieutenant Harold Tittman described how several pilots lost their lives due to material failure: "A parachute was often opened in the air [for target practice] and we used to shoot at it from our planes while performing acrobatics around it. It was during one of these maneuvers that Lt Hagadorn . . . was killed when the wings of his plane dropped off while in the air."[60] One could hardly blame Lieutenant Hagadorn. Nevertheless, the majority of mishaps did receive the label of *pilot error.*[61]

Lieutenant Roscoe's deadly sneeze and the hotshot pilot in the P-40 demonstrated flight's intolerance of airmen's mistakes. The National Advisory Committee for Aeronautics (NACA) analyzed the records of 1,434 accidents that occurred before 1929 and found that 49 percent were due to pilot error. This seemed low compared to subsequent data. An Air Corps survey of accidents in 1932, for example, blamed 65 percent of all accidents on airmen.[62] This rate matched the findings of the National Research Council's Fifth Crash Injury Conference in 1945. The conference reported that of the 5,710 aircraft accidents that occurred within the Army Air Forces in the United States from January through August 1945, two-thirds resulted from "pilot error."[63]

In trying to assess the nature of aircraft accidents that occurred in 1918 and 1919, the Army Air Service determined that 45 percent resulted from "bad landing," 12.8 percent from spins, 9.6 percent from air-to-air collisions, and 7.8 percent from engine problems and other miscellaneous causes.[64] The statistics indicated the prevalence of pilot error because bungled landings, unrecovered spins, and collisions with other aircraft incriminated the operator more than the machine. But these categories were of little use in finding the real causes of a mishap. In 1928 the assistant secretaries for aeronautics from the War, Navy, and Commerce Departments asked NACA to evaluate aircraft accidents and develop a method to classify and understand their causes. NACA formed the Special Committee on the Nomenclature, Subdivision, and Classification of Aircraft Accidents and directed it to root out and catalog the causes of aviation disasters.[65]

The NACA committee's philosophical position was that all accidents were due to one or more definable, knowable faults: either the operator or the machine was to blame, sometimes both. It acknowledged that all flaws originated from some sort of human error, but to distinguish between human and material defects, it limited the consideration of human error to personnel directly involved with the operation of the aircraft.[66] Concerned with identifying exactly which part(s) of the technological system failed, the committee portrayed the pilot as one of many mechanical components, "the actual manipulator of the controls." Described in these terms, the pilot was yet another component susceptible to a variety of malfunctions. Accidents caused by some failure of the human component were divided into five categories: "error of judgment," "poor technique," "disobedience of orders," "carelessness or negligence," and a potpourri category of "miscellaneous errors."[67]

The vast collection of aviation mishaps showcased specimens from each category of pilot error. Lieutenant Frank Flynn provided a fitting example of "error of judgment" on 16 September 1928 when he tried to land in an unfamiliar hayfield at night and struck a telephone pole, tearing the wing off his airplane.[68] Apparently Lieutenant Flynn could have chosen to land elsewhere since the Air Corps blamed 100% of the accident on him. His episode fit NACA's definition of an error in judgment as any imperfect decision on the part of the pilot. "Poor technique" applied to accidents caused by a lack of skill. Disorientation in inclement weather due to inferior instrument flight skills, the inability to recover from a spin, or an inadvertent stall could earn this label.[69] "Disobedience of orders" involved blatant disregard of rules and standards, as evidenced in Lieutenant Roland Birnn's account of the reckless young hotshot who attempted an aerobatic maneuver below minimum safe altitudes and smashed into the ground. The category of "carelessness or negligence" covered a wide variety of errors, including Lieutenant Roscoe's sneeze incident, as well as the case of a pilot who became disoriented and plowed into an airport boundary marker during an attempted takeoff.[70]

NACA's aircraft accident committee considered these types of error as *immediate causes* of aviation mishaps. Under this system of analysis, the accident was directly caused by some manifestation of poor technique, disobedience, carelessness, or the like. But identifying the immediate cause was not enough. A plane may have crashed because of a pilot's bad judgment or carelessness, but why did he or she err or become careless in the first place? The committee established that investigators must also seek the *underlying causes* of the accident.[71] The growing record of aircraft accidents suggested two broad categories of underlying causes: "lack of experience" and "physical and psychological causes."[72] The first category permitted clear definition and included total and recent flight experience. The second one used relatively vague terms that allowed subjective interpretation and required investigators to delve into the inner workings of the human component.

According to NACA's classification scheme, the underlying "physical and psychological causes" of an accident involved either chronic or temporary defects in body and mind. Physical defects included various ailments such as vision problems, hypoxia, decompression sickness, fatigue, and epilepsy. Did Lieutenant Flynn strike a telephone pole because he had the bad judgment of trying to land at night in a strange field, or was it really because he was tired or had a temporary defect in his vision? A psychological issue such as mental fatigue or some form of neurosis or even psychosis may also have played a part. To explain slow or mistaken actions during flight, NACA created a psychological condition called "poor reaction."[73] Poor reaction resembled what one Royal Naval Air Service physician described in 1918 as "brain fatigue," or a mental fog where the pilot loses his normal ability to think and act quickly.[74] Failure to recognize a dangerous landing area, for example, could be due to some sort of cognitive shortcoming that stymied swift, appropriate interpretation or action.

The malleable and indistinct terms associated with underlying physical and psychological causes of accidents enabled investigators to fix blame. If a pilot made a blatantly foolish error, perhaps an official inquest could be satisfied with a diagnosis of a physical ailment, poor reaction time, or some temporary mental lapse due to fatigue. As Peter Galison noted in his analysis of jet airliner accidents, investigators gravitated toward localizing the cause to a specific component in the technological system. Finding a specific cause permitted a satisfactory interpretation of events and empowered remedial measures to help prevent similar mishaps in the future.[75] Modern accident investigations, therefore, may echo the focus on causal determination established by NACA in 1928. NACA reaffirmed this approach in 1930 with a follow-up report on analytical methods for aircraft accidents. The 1930 report noted that the Army Air Corps and the US Navy both adopted this system.[76]

An important aspect of the NACA-led approach to accident investigations involved the attribution of error to multiple sources. The committee acknowledged that an accident might result from more than one cause: "Where two or more factors cause an accident, part will be charged to each."[77] A 1934 accident involving Captain Burton Lewis illustrated how the Air Corps applied the NACA system. On a routine flight in a Douglas BT-2B biplane, Captain Burton encountered fog conditions that required an immediate landing. He found a suitable pasture beneath him, but while maneuvering to land he allowed the aircraft to exceed a safe bank angle and the wingtip struck the ground. The aircraft was destroyed, but Captain Burton survived. The Air Corps' Aircraft Accident Classification Committee split the blame for the cause of the accident between "poor weather" (50 percent) and "error of judgment" (50 percent). The committee also noted Captain Burton's recent lack of experience in the BT-2B as an underlying cause.[78] Affixing half of the blame on the pilot and noting the crash was aggravated by inexperience allowed the investigators to suggest that future accidents could be avoided if pilots maintained better proficiency.

In Captain Burton's case, the crash resulted from a combination of weather and human error. Similarly, NACA determined that blame for a crash could be spread among a system's different components. If a pilot crashed after an engine failure because he or she failed to maintain adequate airspeed while gliding down for an emergency landing, then fault could be assigned to the pilot and the "materiel."[79] This reasoning established an important precedent: if accidents could be attributed to a combination of pilot error and poorly constructed material, then it made sense that they could also be attributed to a combination of pilot error and poorly designed material. Put another way, accidents could result from poor design that hindered the optimal interface of the human and machine components.

Designer Error

A P-47 Thunderbolt mishap during a combat mission in the Pacific theater in World War Two illustrates the blurred lines between pilot error and designer error. While on his first combat mission, the pilot noticed abnormal fluctuations in the fuel pressure gauge. This warned of a serious interruption in fuel flow to the engine. He attempted to fix the problem by adjusting the fuel and engine controls, and his formation leader instructed him to toggle his fuel selector switch from the main tank to the fully fueled reserve tank to ensure an adequate supply of gas. The fluctuations persisted, and eight miles from the home base in Borneo the engine quit. To avoid parachuting into the jungle, the pilot attempted a belly landing on a sand bar adjacent to a large river. He misjudged the landing, and the P-47 stalled and struck its tail on a steep part of the river bank. The aircraft remained intact but careened into a deep part of the river, drowning the pilot.[80]

The investigators dragged the wreckage from the water and noted that the fuel selector switch had been moved toward the reserve fuel tank. Unfortunately, the pilot had not positioned the lever far enough to the side to activate the mechanism and allow fuel to flow from the reserve tank. Although the switch appeared at a glance to be in the correct position, it was little more than halfway over, causing the engine to cut out due to fuel starvation. The accident investigator attributed the accident to only one cause: pilot error.[81] He also criticized the pilot's carelessness since the engine would probably not have stalled with the switch in the correct position. Another pilot challenged this ruling: "It seems to me under the circumstances the pilot was not totally to blame. He was upset and excited because this was his first mission. He thought he turned the selector switch all the way. . . . The selector switch should be designed so that there can be no possible way to switch it part of the way."[82] In this pilot's view, the switch's designers needed to anticipate imperfect action by pilots and reduce the opportunity for human error. In modern language, the system needed to be user-friendly and foolproof.

From his work in human factors engineering at the Aero Medical Laboratory's Psychology Branch after 1945, psychologist Walter F. Grether argued that improper design "trapped pilots into making fatal mistakes."[83] The P-47 pilot was ensnared by a poorly designed switch, and his case was one of many where cockpit engineering invited critical blunders. A Psychology Branch study from late 1945 labeled the improper movement of a control or switch as an "adjustment error." These accounted for 18 percent of pilot errors associated with the operation of cockpit equipment. "Substitution errors" were to blame 50 percent of the time.[84] Unlike the adjustment error that took place with the P-47 fuel switch, a substitution error resulted when an airman confused one control for another. This type of error was illustrated by the copilot of a fully fueled, bomb-laden B-25 Mitchell bomber shortly after takeoff. According to the pilot, "We crossed the end of the runway at an altitude of two feet and were pulling up over the trees shortly ahead when I gave the 'wheels up' signal. The airplane mushed [lost altitude] and continued to brush the tree tops at a constant 125 mph speed. . . . The co-pilot had pulled up the flaps instead of the wheels."[85] The levers for the landing gear and flaps were located right next to each other, and they were down near the floor between the pilot and copilot seats. This required the pilot or copilot to turn his head and look down, a risky proposition during critical phases of flight. Actuating them by mere feel enabled flyers to keep their gaze forward, but such design also made for an easy substitution error and possible disaster.

Associating accidents with equipment design rather than just human error marked a new perspective in the man–machine relationship. As described in previous chapters, improvements in aircraft performance during the interwar years required aircraft designers to confront the problems of human limitations. Engineers collaborated with life scientists to bridge the widening gap between machine and human capabilities, and this effort altered the relationship between airman and aircraft. Overall system performance still suffered, though, from a poor functional interface between operator and equipment. It was an inefficient cybernetic system. The Psychology Branch's study of the link between equipment design and pilot error confirmed the hazards of inefficient design and the need for greater attention to how airmen were integrated into a complex technological system. Design modifications could diminish "pilot error factors" and thus help avoid crashes in the first place.[86] Yet design improvements were not enough. Enhancing the safety and efficiency of the system required careful engineering of the airman's thoughts and actions as well.

Meat Sprockets: Standardizing the Human Component

During World War Two, the Army Air Forces struggled to balance quality and quantity. High-altitude daylight precision bombing, the heart of US airpower strategy, required highly sophisticated manned bombers. Airmen needed extensive training to operate these complex machines in environmental extremes, penetrate enemy defenses, find the target, and return to distant airbases in unpredictable weather conditions.[87] The intensive bombing campaign and heavy attrition in the European theater in the years before the Normandy invasion increased the demand for high quantities of combat airmen, yet this reduced the quality of the human component since adequately trained and experienced air crews were in short supply. A key figure in the strategic bombing campaign against Germany noted this problem. In January 1943, a month before taking over as the commander of the Eighth Air Force in England, Major General Ira Eaker wrote to Major General George Stratemeyer, the chief of the Air Staff.[88] A military aviator since 1918, General Eaker lamented the lack of training and experience in the current crop of airmen:

> The one thing which has been the greatest shock to me in the 5 months experience I have had with our bombers in this theater, is the difference in capacity and qualification between the wartime crews and the old peacetime standards we used to have in the First Pursuit Group, the Second Bombardment Group and the Third Attack Group. That affects everything. These green kids, bless their hearts, make mistakes you and I would never have dreamed of with all the experience we had in peacetime. . . . I remember that we formerly would not let an officer fly a four-engine bomber who had not had 10 years service, and here we have youngsters with 10 months' service serving as crew captains. Instead of raving and ranting at them for their inexperience, what we have to do is to revise and recut our operational plans to fit their level of experience. Of course, we do everything we can to pound into them the things they must know in order to do a satisfactory job. I feel we have made a lot of progress on this point.[89]

General Eaker's concerns were echoed a year later by another Army Air Forces' general who noted that the capabilities of combat personnel were at a bare minimum, and that "we have been forced to sacrifice quality for quantity."[90]

In order to obtain consistent performance of the highest possible quality from variable and inexperienced human components, the air force sought to standardize aircrew actions to the maximum extent. This involved universal training methods that equipped airmen with the basic skills necessary for success in combat. It also included the use of checklists that guided and standardized an airman's activities in various phases of flight. Achieving reliable, efficient performance also prompted more attention to cockpit design. Overall, these efforts treated the airman as an automaton, an organic mechanism that operated in a programmed, predictable, machine-like fashion that lessened the chance for error and compensated for inexperience. In this reductionist view, a pilot, navigator, gunner, or bombardier was a servomechanism, a cog in a machine, an interchangeable part, an organic subcomponent—a meat sprocket—which could achieve a baseline standard of performance when inserted into the system.[91]

If a bomber and its ten-man crew are seen as a complex cybernetic system, training the human components to function as reliable "manual servomechanisms" enhanced their ability to process and apply information in an efficient manner.[92] This required a streamlined, standardized training program for airmen that could shape their actions and "pound into them the things they must know." Such pounding molded them into one component of many, reducing them to important yet unwieldy and often unreliable organic sprockets in a machine that, if it failed, would likely destroy them.

In their 1936 book, *This Flying Game,* Ira Eaker and Henry "Hap" Arnold, who would become the commanding general of the US Army Air Forces from 1941 to 1945, noted that combat operations during World War One demanded the development and uniform application of specific training methods.[93] No wonder, then, that Eaker brooded over the minimal training received by his bomber crews in the Second World War. Eaker and Arnold maintained that proper training taught pilots to perform their duties instinctively and with little thought or effort.[94] This was especially important for critical tasks involving aircraft control or proper operation of life-support equipment.[95] A 1942 training manual emphasized the mastery of basic flight procedures and the development of a reliable "mental process" that prevented bad habits or mental confusion from seeping in.[96] Effective training made good habits second nature in as short a time as possible and was essential for the cybernetic system to function properly in the dynamic environment of flight.

The importance of training airmen to behave in a predictable, efficient, machine-like manner was illustrated by two B-29 mishaps that occurred on the same day in 1944. Each bomber experienced an engine fire and the loss of two of its four engines, but these similar problems resulted in markedly different outcomes. In the aircraft that survived, the pilot commanded the engineer to apply the step-by-step procedure to extinguish an engine fire and shut down the two affected engines. As a precaution after the fires were extinguished, the pilot ordered the rest of the crew to bail out in an orderly fashion. The pilot stayed with the aircraft and made a successful landing, and there were no injuries among the personnel who parachuted to the ground. This crew operated correctly and complied with its training in a textbook manner. The crew in the other B-29, however, met disaster due to a malfunction of the human component.

In the doomed B-29, after one engine failed and another caught fire the crew failed to apply the emergency procedures to extinguish the engine fire. As the fire raged on, the copilot panicked and "froze" at his controls; in the meantime, the pilot left the cockpit and clamored for his parachute. Other crewmembers scrambled for their parachutes as the aircraft rolled out of control. Six of the crew bailed out. One had his parachute on upside down; another was killed due to insufficient time for the parachute to open. Nine crewmen failed to jump out and died on impact.[97]

The B-29 mishaps demonstrated the importance of adequate training that permitted a functional, reliable interface between man and machine. Closely associated with aircrew training was the concept of *standardization*. Uniform performance of cybernetic systems required standard, unfailing components. To achieve consistent performance among human components, the Army Air Forces directed instructors at training bases to produce airmen "who think along the same lines and speak in the same terms."[98] This anticipated Norbert Wiener's 1948 observation that effective incorporation of human components into a cybernetic system required their function to be quantifiable and predictable.[99] Airmen who showed up in combat units with the same skills as other airmen enabled a more efficient application of the air weapon. Standardization, therefore, helped to overcome a lack of experience; automata worked the same way every time. Viewed through the lens of human factors engineering, the integration of human and machine also involved standardizing these system components.

Standardization involved rote, consistent training as well as human factors issues such as cockpit design. One aspect of design relevant to standardization was the placement of controls and instruments in the cockpit. Different types of airplanes placed similar controls in different locations, and this caused considerable confusion and error. In World War Two aircraft, such as the B-25 and the C-47, for instance, three different levers labeled "propeller," "mixture," and "throttle" controlled each engine. These adjacent levers allowed the pilot to adjust engine performance during different phases of flight. As one could surmise, these levers invited confusion and required training and practice in proper use. In the C-47, the levers were placed next to each other in this order: propeller, throttle, mixture. The order was different in the B-25: throttle, propeller, mixture. This spelled trouble for pilots transitioning between the C-47 and B-25.[100] A pilot with experience in both aircraft was prone to accidentally moving the propeller lever when he wanted to move the throttle lever instead. Of course, this could cause a significant, unwanted change of power during a critical phase of flight.

The standardization of cockpit engine controls promised to reduce the likelihood of pilot error, and several organizations pushed for a consistent arrangement of engine controls in different types of aircraft. Operating under the premise that "accidents do not happen, they are caused," the Army Air Forces Office of Flying Safety collaborated in 1944 with the National Research Council and the American Society of Mechanical Engineers to study the proper placement and standardization of engine controls as well as other cockpit instruments. In addition to reducing confusion, a uniform, standard design of the cockpit environment enabled more efficient training. Good habits in one airplane could be bad habits in another if the controls were reversed. Airmen could gain proficiency in less time and perform their tasks in a more reliable manner if the cockpit environment remained as consistent as possible.

The quality of aircrew performance was also improved by a simple, effective form of standardization: the checklist. Like a recipe, a checklist consisted of written, step-by-step procedures that ensured airmen performed their duties in the correct manner and sequence. Even experienced pilots benefited from this tool. Consider the tragic event that convinced the Army Air Corps to institute checklists in aircraft operations. On 30 October 1935, the Air Corps wheeled out the prototype of Boeing's B-17 Flying Fortress at Wright Field, Ohio, for a demonstration flight. The test pilots taxied the bomber onto the runway, pushed the controls for all four engines to full power, and the shiny new bomber lumbered forward. As it continued to gain airspeed and climb away from the ground, the nose began to pitch upward at an unexpected rate. The pilots pushed forward on the control column to lower the nose, but with no effect. With its engines at full power, the bomber entered an alarmingly steep climb and stalled a few hundred feet above the ground. It performed a graceful half-pirouette and pointed straight back to the ground. The nose started to pitch up again as the aircraft gained speed in its downward plummet, and the bomber slammed onto its belly, bursting into flames. Two of the five crewmembers died.[101] The cause: pilot error.

The copilot, Lieutenant Donald Putt, recounted what happened: "We took off with the controls [elevator and rudder] locked. It was the first airplane in which you could lock the control surfaces from the cockpit. When we taxied out, for some reason we were in a hurry. Those were the days before the checklists."[102] Control locks prevented the bomber's large elevator and rudder from flailing about during high winds on the ground, and Boeing engineers had installed a special lever in the cockpit so the pilot could unlock the controls. The pilot forgot. The copilot forgot. Boeing's chief test pilot, riding along as an observer, forgot.[103] Although the crewmembers were highly experienced, they had little time to develop the important habits and procedures necessary for this specific aircraft. The Air Corps applied the hard lesson of this tragedy and ordered the use of checklists during aircraft operation. The official B-17 flight manual in World War Two cautioned pilots that operation of the B-17 was too complex for even experienced pilots to memorize; the checklist was the "only sure safeguard" against pilot error, and it was "absolutely essential that the cockpit checklist be used properly by pilot and copilot at all times."[104]

Checklist usage guided both the thinking and physical action of airmen. A 1943 manual stipulated that all pilots must use a checklist before takeoff. In large aircraft with a pilot and copilot, checklist discipline also required the pilot to verbalize each step of operation and await reply from the copilot.[105] In the B-29 Superfortress's Before Take-Off Checklist, for example, the pilot called out each checklist item, and the copilot responded: "Emergency Brakes" . . . "Checked;" "Wing Flaps" . . . "25 Degrees;" "Autopilot" . . . "Off;" "Propellers" . . . "High RPM."[106] These were only four of the four dozen checklist items performed by the pilot before becoming airborne. Moreover, the flight engineer, navigator, bombardier, radio operator, and gunners all had their own checklists to complete for each phase of aircraft operation. In essence, checklists programmed an airman's movement and function. They imprinted machine logic on human action. Pilots, of course, still possessed ultimate agency over aircraft, and military aviation benefited from human judgment and flexibility. Nevertheless, efficient operation of aircraft required a rigorously disciplined sequence of movements performed in the same way, every time, in rote fashion, and by automaton-like airmen.

Standardizing the human component to ensure reliable integration with the rest of the system anticipated the cybernetic model, yet it also invited comparison to another concept involving human efficiency: Taylorism. In many respects, the efforts to make airmen more efficient reflected the principles of scientific management put forth by Frederick Winslow Taylor in 1911. Opti-

Lieutenant Henry "Hap" Arnold, future five-star leader of the US Army Air Forces in World War Two, in 1911. Major Arnold's 1925 report, "The Performance of Future Airplanes," traced the rapid improvements in aviation capabilities since 1911 and predicted the need for advanced life-support systems. Photograph courtesy of the US Air Force.

The Kettering Bug. A collaboration of Orville Wright, Elmer Sperry, and Charles Kettering resulted in the Kettering Bug. Flight tested in 1918, it was the Army Air Service's first gyroscopically stabilized unmanned bomber. Photograph courtesy of the National Museum of the US Air Force.

Sperry Gyroscope Company's artificial horizon indicator of the type used by Lieutenant Jimmy Doolittle in the first zero-visibility flight and landing in 1929. This device compensated for the human inability to maintain orientation with the Earth's horizon and fundamentally advanced aviation capability. Photo by Eric Long, Smithsonian National Air and Space Museum (NASM 2014-04796).

Test pilot Wiley Post's automatic pilot, "Mechanical Mike," built by the Sperry Gyroscope Company and used in Post's 1933 around-the-world flight. The *New York Times* described this as Post's "robot pilot" that let him "sit relaxed in his chair, an overstuffed one specially installed, without much regard for the stick technique" and "held the plane on a set course without aid from the pilot." Photo by Eric Long, Smithsonian National Air and Space Museum (NASM 2012-01250).

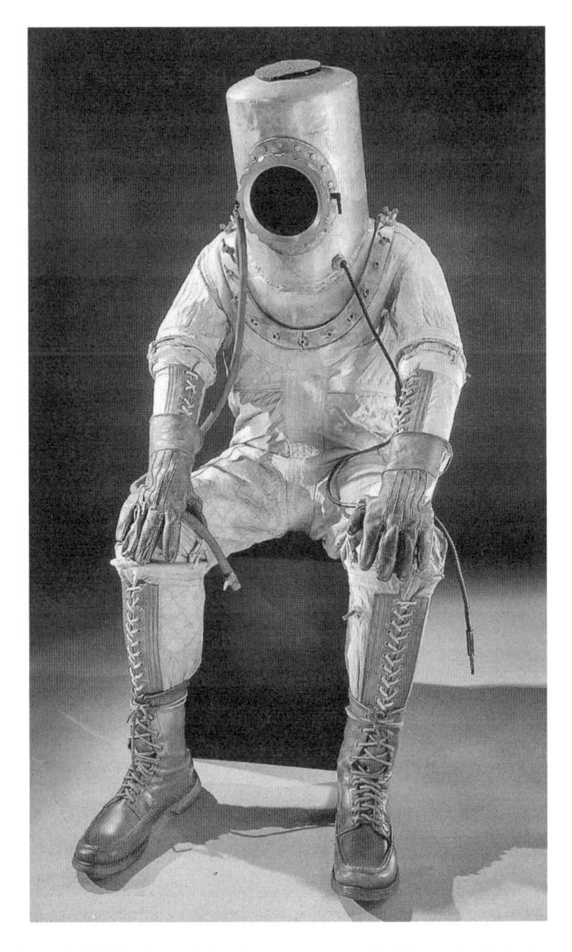

Test pilot Wiley Post's 1935 high-altitude pressure suit. Photo by Eric Long, Smithsonian National Air and Space Museum (NASM 98-15012).

Aviation medicine pioneers Walter M. Boothby, William R. Lovelace II, and Harry G. Armstrong receiving the Collier Trophy, aviation's highest award, from President Franklin D. Roosevelt in 1939 for their contributions to aircrew performance and safety. Photo courtesy of the Smithsonian National Air and Space Museum (NASM A-48076-L).

Cockpit of Boeing B-24 Liberator. This image shows the significant proliferation of cockpit instrumentation that required pilots to become information managers heavily augmented by an artificial representation of reality. Unmanned, remote-controlled variants of B-24s and B-17s were also used during the Second World War. Photograph courtesy of the National Museum of the US Air Force.

This unique perspective on a World War Two B-26C illustrates the aircrew before donning oxygen masks. The airman in the nose of the aircraft is capitalizing on the opportunity for a cigarette before the aircraft enters the extremes of high altitude and combat. Photograph courtesy of the USAF Historical Research Agency.

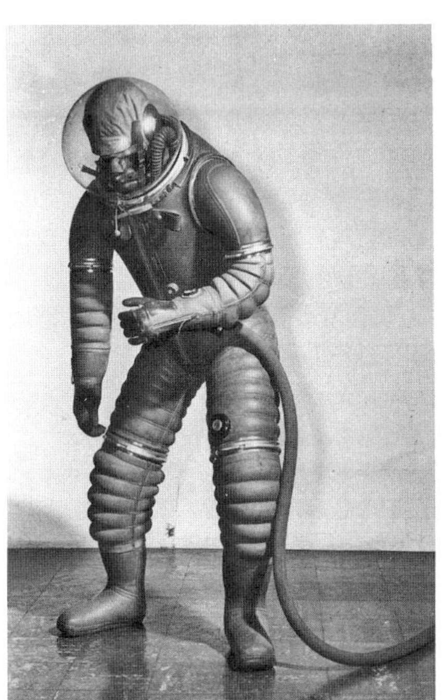

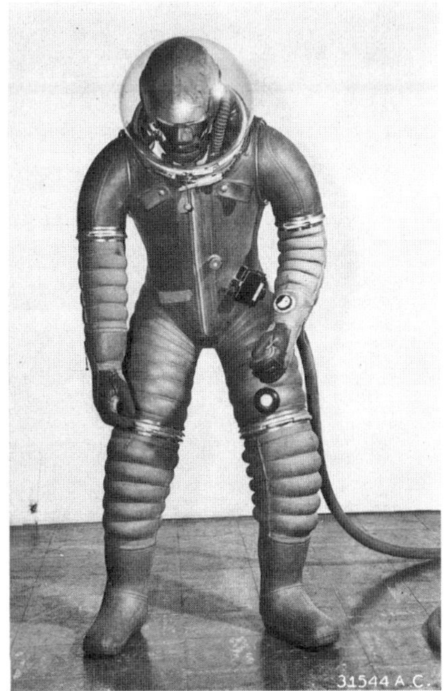

Goodrich pressure suit, circa 1943–50. This World War Two–era pressure suit prototype illustrates the necessity of enabling pilots to function in increasingly austere environments. Photo courtesy of the Smithsonian National Air and Space Museum (NASM USAF-31544AC).

Modern pilot helmet from the F-35. The $400,000 F-35 helmet portrays a computer-adjudicated view of the environment and enables the pilot to "look through the jet's eyeballs to see the world as the jet sees the world." Photograph courtesy of the US Air Force.

mizing human performance required careful management and control of each physical action, and Taylor treated workers as system components that could be finely tuned to increase overall production. In one of Taylor's examples, pig iron laborers could be managed or programmed like automata to perform their duties in a manner that integrated them into the system in the most efficient manner possible. In one sense, pilots flying aircraft resembled laborers handling pig iron: both were governed by Taylor's notion that "every single act of every workman can be reduced to a science."[107] Such standardization of aircrews through training programs and checklist discipline paralleled the efforts of scientific management to develop "rules, laws, and formulae" that guided the actions of individual workers.[108]

Although Taylorism permeated industrial practice in the early twentieth century, there is little direct evidence that it consciously influenced the community of practitioners in military aviation. Yet notable similarities existed between scientific management and the technological changes in flight. Perhaps unwittingly, the air force applied Taylor's principles from 1911 and ended up developing what Wiener described in 1948 as cybernetics. Taylor observed that "in the past the man has been first; in the future the system must be first," and Wiener noted that cybernetics was a new science that "embraces technical developments with great possibilities for good and for evil."[109] The evolution of aviation technology reflected both statements: the air force altered man's relationship to technology in order to optimize the aircraft as both a tool and a weapon system.

The effort to balance quality and quantity of personnel in Army aviation shaped the airman's role in flight. Even after passing through the selection filters established by flight surgeons, individual airmen varied in their physical and mental capabilities. In order to mold them into reliable components of the airman–aircraft complex, the air force trained them to operate in similar, predictable ways and to submit to the authority of checklist procedures. That way, they could be inserted into combat operations with some degree of confidence in their ability to perform the mission. Just as Taylorism aided the ascendancy of the manager, military aviation's quest for quality and efficiency elevated the designer and engineer.[110] As flight became more and more the management of a complex technological system, there was a shift in authority from the individual airman to engineers who designed cockpit systems and managers who constructed training programs and checklists. Enhancing efficiency and minimizing error also involved a greater reliance on automation.

The Automation Trajectory

In 1937, the assistant chief of the Air Corps, Brigadier General Hap Arnold, lamented that a disturbingly high percentage of aircraft accidents resulted from human error. This provided sufficient reason, he argued, to "relegate the human flyer and elevate the mechanical pilot."[111] The Air Corps saw an array of benefits in automating flight. In addition to decreasing the opportunity for pilot error, automation promised to reduce the time required to train competent airmen. Automation also eased pilot fatigue and improved the ability to fly and navigate through inclement weather. Airpower enthusiasts, particularly the mavens of strategic bombing, highlighted automation's ability to enhance the accuracy of strikes from high altitude. Efforts to automate more and more aspects of flight represented a "technological trajectory," or a distinct pattern of change, within the complex system of military aviation.[112]

The trajectory toward automation began even before Army aviation made its debut in the Great War.[113] In 1912 famed inventor Elmer Sperry began adapting his maritime gyroscopes to the even more difficult task of stabilizing vessels in the air.[114] Sperry was aware of the need to improve the flying qualities of the notoriously unstable aircraft of the day. In 1913 he and his son had installed a gyroscopic-controlled stabilizer on one of the Navy's Curtiss flying boats.[115] They demonstrated that aircraft could be flown safely, smoothly, and reliably while under mechanical control.[116] Eager to enhance the stability of aircraft, the Signal Corps expressed interest in Sperry's automatic stabilization device in 1916.[117] This interest failed to materialize into widespread operational use during the First World War, but it became an integral aspect of flight during the interwar years.

Improvements in autopilot technology occurred in parallel with developments in instrument flight during the late 1920s and 1930s. The Guggenheim Fund, a major locus of instrument innovation in the late 1920s, reported that the traditional tasks of controlling pitch, roll, yaw, and altitude could be delegated to a machine, thus relieving the pilot from constant monitoring of the altimeter and allowing more attention to navigation and other duties.[118] Army lieutenant and instrument pioneer Albert Hegenberger worked with the Sperry Gyroscope Company in 1929 to test the gyroscopically powered autopilot's usefulness as an aid to the pilot during instrument flight. Dubbed the "Mecaviator," short for "mechanical aviator," this device used information feedback and control to fly the aircraft with greater precision than a human pilot.[119] In his correspondence with airpower icon William "Billy" Mitchell, Hegenberger noted that while instrument flight fatigued the pilot, an automatic pilot could take over and fly the aircraft "as well or better" than the pilot and for a longer period of time.[120] Although the name failed to catch on, the Mecaviator symbolized the rise of mechanical control. It also portended Norbert Wiener's effort two decades later to develop "a mechanico-electrical system" that could "usurp a specifically human function."[121] Hereafter, flight could be performed by mechanical devices, while the pilot observed the cockpit instruments and managed the status of the mechanical components.

By the early 1930s, the demand for long-duration flights through fair and foul weather helped convert the autopilot from a stabilizing mechanism to a much more elaborate device. In its primitive form the autopilot emulated the human sense of balance, reacting to the inertial changes sensed by a gyroscope much like the human body reacted to changes in rotational velocity detected by the semicircular canals in the inner ear. Improvements in the science and technology of gyroscopes, instruments, and electronics further enabled autopilots to gather inputs from flight instruments and radio signals and translate them into a more sophisticated form of aircraft control. This improved the capabilities of automatic control systems just as vision and hearing enhance a human's ability to interpret the environment. Treating aircraft as cybernetic systems, these advances anticipated Norbert Wiener's observation that such systems could perform progressively more elaborate, complex tasks. Increasingly accurate performance marked the trajectory toward greater reliance on automated flight in instrument conditions.

Work in the Air Corps' Materiel Division at Wright Field helped fuel the trajectory toward fuller automation. In 1931 the division outfitted a Fokker C-7A aircraft with a Sperry "Gyro, Type A-1, Pilot."[122] Tests that year demonstrated how the Sperry component could cooperate with the aircraft compass to maintain a desired heading, and the Air Corps predicted that the device would soon be capable of using information from a radio beacon to fly the aircraft on a specified course.[123] This prediction came true. By 1935 the Sperry autopilot improved its "elaborateness of performance" by coupling a directional gyro with an automatic turn-control feature, thus enabling the autopilot to place the aircraft into a turn and maintain the turn. It could also interpret data from a radio compass to follow a desired course.[124] This pattern of development led to an automation-related breakthrough in 1937.

A new capability emerged from the interconnection of a gyro-stabilized autopilot, radio marker beacons, and a radio-sensitive directional compass: aircraft could fly *and navigate* without direct input from a pilot. On 23 August 1937, Air Corps test pilot Captain George V. Holloman took off in his Fokker C-14, engaged the automatic controls, and then sat back and observed as the aircraft followed radio signals back toward the runway and landed. The Air Corps trumpeted this fully automatic landing as "one of the most remarkable accomplishments in all the annals of aviation."[125] It was a far-reaching achievement—but for engineers, not pilots. Captain Carl J. Crane, the director of the Instrument and Navigation Unit at Wright Field and one of the primary engineers of automated landings, noted that substituting automatic for manual control reduced the pilot to a monitor of high-tech gizmos—"a light task which reminds us of that gag where one indicator among very many others is pointed to as the one that tells the pilot when all the others are working."[126] Despite infringements on the pilot's traditional role, Captain Holloman touted greater automation as a positive development. Since modern cockpits were being overrun with instruments, switches, and assorted devices, all of them demanding the pilot's attention, he believed that delegating workload to automatic devices helped the pilot avoid distractions and become a more efficient manager.[127]

A major motive for developing automated, hands-off landing technology was to hasten the training of airmen during a national emergency.[128] Hand-flown landings in poor weather required lengthy training and practice in instrument procedures, and devices that augmented human skill could help airmen achieve an acceptable level of proficiency in less time. Although fully automated landings remained a rarity in the late 1930s and through World War Two, the air force pursued other practical applications of automation technology to enhance safety and ease the training burden. The inclusion of autopilot technology in pursuit (fighter) aircraft, for example, promised to reduce pilot error during instrument flight and make up for deficiencies in training.

One of the chief proponents of installing autopilots in pursuit aircraft was General Delos C. Emmons, the commanding general of Headquarters Air Force at Langley Field, Virginia. In July 1939 he recommended equipping single-seat aircraft with a so-called Junior Automatic Pilot. This device provided special value in fighter aircraft, he argued, since the single pilot lacked the aid of another pilot and was more prone to spatial disorientation.[129] General Emmons's views were confirmed by a special board convened to study the subject of instrument flight and related accidents in fighter aircraft. It concluded in 1939 that such inherently unstable aircraft demanded excessive concentration by pilots, especially inexperienced ones. These problems could be attenuated by automatic controls that relieved "mental and physical strain."[130] Nearly two years later, as the threat of war increased, General Emmons expressed his concerns to the chief of the Air Corps regarding pilot ability:

> There is no question but what we can train our pilots to fly the present types of pursuit planes on instruments *but such training requires time*. We are going to be forced to use pilots in pursuit airplanes who will be unable to fly these airplanes on instruments without a very high casualty rate because we will not have the time to give them the necessary training. . . . The Junior Automatic Pilot not only will solve to a large extent the matter of flying on instruments but it is a very useful device in other respects. It permits the pilot to devote his attention to the adjustments of engine controls, permitting much better performance, and is a very definite aid to navigation. . . . The duties of the pursuit pilot are constantly increasing and this is the one invention that helps him take care of this increasing load.[131]

In his effort to advance the promise of airpower, General Emmons essentially argued that installing a (superior) machine substitute may overcome the poor quality of the human component.

The development of automated flight-control technology signaled a pattern of change in military aviation, a pattern in which airmen submitted more and more of their role to machine augmentation. The air force shifted the complex man–machine system onto a trajectory toward automation in order to overcome various human limitations and enhance the potency of the air weapon. This was particularly crucial to supporting airpower's core strategy of high-altitude daylight precision bombing. *Precision* required sophisticated electromechanical devices to guide aircraft on their bomb runs and release their payloads at exactly the right time. This strategy depended on weapons of extremely high quality, and the optimal employment of such weapons required airmen skilled in their proper use. The Eighth Air Force's bombing raid on Vegesack, Germany, illustrated how the system achieved maximum effectiveness.

On 18 March 1943, more than 100 B-17 and B-24 bombers struck the submarine pens at Vegesack in northern Germany. This was the largest daylight bombing raid against the Reich so far, and the bombing accuracy was so precise that Winston Churchill sent a congratulatory note to the commander of the Eighth Air Force, General Ira Eaker.[132] An official account from the Eighth provided a candid summary of the raid: "We had a rather remarkable bombing mission . . . the remarkable part of the mission being that the boys hit the point they were aiming at, and that the mission was a very great success."[133] Pleasant surprise that the boys actually hit their target was not exactly a ringing endorsement of high-altitude daylight precision bombing; nevertheless, the Eighth attributed the raid's success to the fact that it was the first large-scale use of Automatic Flight Control Equipment (AFCE).

Tests in early 1941 verified that combining AFCE with the bombsight improved bombing accuracy to a significant degree, at least under controlled, noncombat conditions.[134] During operational missions prior to Vegesack, however, mechanical problems plagued the delicate AFCE, and pilots chose manual control during the critical phase of the bombing run. Concerned that precision bombing could not reach its full potential under manual control, the Materiel Command assigned a group of bombardier instructors to evaluate these problems. The group verified the efficacy of automatic control and recommended key modifications that improved the system's reliability. Tests of the modified system showed significant improvements, and the lead aircraft in the Vegesack raid employed the improved AFCE to notable effect.[135]

A spokesman from the Eighth briefed other organizations on why the Vegesack bombs hit their mark. "We have known for some time that bombing results have been unsatisfactory," he explained, "because the airplane hasn't been accurately operated to get the desired degree of accuracy."[136] The airplane, in the Eighth Air Force's view, worked fine; it was just operated incorrectly. Maximum effectiveness demanded that aircrews switch to automatic control—the human component had to defer to the mechanical component. This required the pilot to relinquish control to the bombardier during the final run to the target. The bombardier, in turn, fine-tuned the AFCE and the bombsight to fly the most precise approach to the target.[137] Overall, precision bombing required the airman to be "consistent in technique, and not affected by fighter interference, or by flak bursting around him."[138] In other words, airmen were supposed to interact with the rest of the system by following strict procedures, activating the correct sequence of switches, and turning control over to other, more precise components that operated undisturbed by fighters, flak, or fear.[139]

Other reports from American bombing campaigns in the Second World War highlighted the vital link between automation technology and the aircrew's willingness and ability to use it. The Ninth Air Force blamed suboptimal bombing accuracy during campaigns in North Africa and the Mediterranean in 1943 not on the quality of the equipment but on insufficient training in how to use and maintain it.[140] This allowed the air force to trumpet the potential of the air weapon despite its disappointing operational results. Strategic bombing enthusiasts attributed errant bombs to lack of proper instruction and practice, or even human error, and claimed accurate bombs resulted from superior technology manned by well-trained, disciplined personnel. A similar phenomenon occurred in the Pacific theater. One squadron commander noted that in 1942 many pilots received little training in how to use the "mechanical brain" and that few were comfortable ceding control to the device when harassed by flak and fighters. Proper instruction in the device's operation and merits, however, enabled aircrews to supposedly improve their bombing accuracy by 25 to 50 percent by the spring of 1943.[141] The measurable difference between manual and automatic bombing suggested that the trajectory toward automation amounted to little unless airmen relinquished control to the automatic component.

Conclusion

Understanding the evolution of aviation technology requires an examination of the functional interaction between airman and aircraft. The concept of cybernetics informs this interaction. Although Norbert Wiener invented the term in 1947, the air force practiced the basic principles of cybernetics in the interwar era and World War Two. Flight relies on a process of information analysis and feedback, and aircraft achieved increasingly sophisticated capabilities due to enhancements in sensors and control mechanisms. In aircraft and other cybernetic systems a poorly integrated, inefficient, or unreliable component impedes system function. Advances in aviation capabilities required careful integration of the human component with the rest of the machine, and the inherently variable human factor had to be rendered more efficient through standardization. Some phases of flight even required substitution of a superior mechanical device.

Human variability introduced innumerable errors and inefficiencies into the system. The thick catalog of aviation accidents illustrated that the majority of mishaps resulted from pilot error, usually in the form of some lapse of judgment or lack of skill. Instituted by NACA in 1928, the model for accident investigation assigned one of the underlying causes of pilot error to physical and psychological failings. This portended future instructions to aircraft designers that they must consider the physical and mental capabilities of airmen in order to maximize aircraft performance. Indeed, aircraft designers compounded the risk of pilot error when they failed to consider how the human component interacted with the machine. In addition to design improvements, another method of enhancing the consistency and quality of the human component was to standardize its function. By emphasizing uniform, reliable procedures, the air force's flight training programs strived to equip airmen with basic skills that helped offset their inexperience. Standardization also mandated the use of checklists to ensure airmen completed the proper tasks in a programmed, machine-like manner.[142] While standardization minimized the inherent variables of the human component and reduced the risk of pilot error, the trajectory toward greater automation enhanced the performance quality of the technological system.

One way to minimize human error and maximize precision was to take control out of the hands of pilots, at least during some flight operations or maneuvers. Lieutenant Colonel Henry's U-2 flew via automatic control while at operational altitudes, relegating him to the role of manager of the aircraft's systems. This change in the pilot's role had long precedent. The 1945 "Airplane Commander Training Manual for the [B-29] Superfortress," for example, instructed the commander that "you are no longer just a pilot. . . . You are flying an 11-man weapon."[143] The Superfortress pilot became a manager and commander of people and equipment, a supervisor of an elaborate cybernetic system.[144] Increasingly complex systems and the trajectory toward automation changed the airman's role in flight. This trajectory further transformed the role of the human component—even before the Computer Age and Space Age—by displacing airmen from aircraft.

Flight without Flyers

At the end of the Second World War, the commanding general of the US Army Air Forces declared: "One year ago we were guiding bombs by television, controlled by a man in a plane fifteen miles away. I think the time is coming when we won't have any men in a bomber."[1] His statement illustrated military aviation's trajectory toward higher performance and greater automation, a trajectory that would alter the man–machine relationship and lead eventually to greater reliance on unmanned aircraft. This vision contradicted the traditional image of pilots as leather-jacketed daredevils in sleek machines defying the elements and vanquishing enemies with their skill, wit, and grit. The crack in this façade deepens when considering how the air force strove to overcome human limitations and expand the capabilities of the machine. To that end, the air force sought to create *pilotless* aircraft since the early decades of aviation.

In 1918 the Army Air Service tested the so-called aerial torpedo, a gyroscopically stabilized unmanned bomber and the ancestor of the cruise missile. Work on the aerial torpedo concept continued throughout the interwar years along with other improvements in aviation technology, such as instruments, autopilots, and navigation aids. By the late 1930s advances in radio, electronics, and automation technology accelerated interest in pilotless aircraft, and the air force developed a variety of unmanned technologies during World War Two, such as remote-controlled heavy bombers and ground-launched cruise missiles. The efforts over these decades not only shaped new roles for airmen but also reflected the air force's institutional interest in gaining dominion over alternative forms of military aviation. Air force leaders sought to expand the scope and legitimacy of the air weapon, even if it required a transformation of the airman's role and a redefinition of flight.

The Aerial Torpedo

Cruise missiles are winged, unmanned, autonomous aircraft that deliver a payload to a specific target. The air force's first cruise missile appeared in World

War One. It was part airplane, part cash register, and part player piano. Airmen dubbed it the "aerial torpedo" since it originated as a project in the US Navy. In April 1916, the Naval Consulting Board promoted the idea of a small, unmanned, gyroscopically controlled airplane that could serve as a flying bomb and detonate hundreds of pounds of explosives on distant targets such as ships or ports.[2] This program was assigned to one of the board's well-known members, Elmer Sperry, resulting in tests of the first unmanned "flying bomb" by early 1918.[3]

Sperry's work on this concept caught the eye of the Army Signal Corps, and in November 1917 the Signal Corps Aviation Section initiated its own aerial torpedo project.[4] The Army contracted with the Dayton Wright Airplane Company, which hired Orville Wright, Elmer Sperry, and another prolific inventor, Charles Kettering.[5] Kettering used pneumatic valves and levers from an Aeolian Player Piano to link Sperry's gyroscope with the controls for the elevator and rudder, and he integrated this apparatus into a small, custom-made aircraft that carried a 185-pound payload. A counting device from National Cash Register determined the number of engine revolutions and cut the ignition after a preprogrammed time. The combined action of these devices enabled the aerial torpedo to maintain a stable flight path and dive toward its target at a predetermined distance.

Nicknamed "Kettering's Bug," or "Bug" for short, the Army's version of the aerial torpedo got off to a rough start. Seconds after its first launch on 2 October 1918 the Bug zoomed upward, rolled right and left, stalled, and plunged into the ground.[6] In the next test three days later, it took off, performed some unexpected aerobatic maneuvers, and then climbed to 10,000 feet and drifted downwind. Kettering and some of his fellow engineers rushed to their cars and gave chase. They recovered it about twenty miles away and, hoping to maintain secrecy, convinced local farmers that the pilot had parachuted out.[7] A third test on 22 October proved successful: Kettering's Bug took off, maintained its course for 500 yards, and dove into the target area at the prescribed distance.[8] The officer in charge of the experiments telegraphed his superiors advising immediate construction of 100 aerial torpedoes, and he urged future deliveries of up to 100,000 units.[9] The initial order was for only 25, however, and the Armistice was signed before the Air Service could let slip the Bugs of war.[10] Nevertheless, interest in the aerial torpedo persisted after the Great War.

The development of the aerial torpedo went through several phases between 1917 and 1932. In the initial phase from 1917 to 1920, the Air Service spent

$275,000 on Kettering's Bug with mixed results.[11] Although some tests met with success, this version of a flying bomb proved too unpredictable. Poor internal construction of the gyroscope caused a high rate of precession or drift that skewed the flight path, and the accelerations imposed during takeoff tended to tumble the gyroscope and send the Bug reeling. Additionally, the pneumatic player piano mechanism that linked the gyroscope to the Bug's directional controls was prone to failure. This assortment of problems impeded the overall progress of the system. In March 1920, the Air Service specified that advancing the concept of the aerial torpedo required a better gyroscope as well as improved control mechanisms.[12] The Air Service also mandated a major modification in the experimental version of the aerial torpedo: room for a pilot.

The manned version of the aerial torpedo was a normal-size aircraft, calling forth a new function for pilots as well as a new phase in the program. The pilot became a technical observer who made occasional adjustments to onboard equipment and used his flying skills only when the automatic controls failed.[13] From 1920 to 1927, Elmer Sperry continued to craft improvements to gyroscopic control. The manned phase of aerial torpedo development also introduced the practice of radio control. Engineers at McCook Field, the Air Service's research and technical training center in Dayton, rigged an automobile with a remote-control device in 1921 "for the purpose of applying it to airplanes later on if it becomes necessary or desirable."[14] It soon became necessary and desirable. Without directional control during flight, the aerial torpedo was little more than an inaccurate form of heavy artillery. If the gyroscope-based controls worked perfectly, it could maintain a preset heading toward the target area, but any amount of wind would blow it off course. In the summer of 1922 engineers integrated a radio-based remote-control device with the aerial torpedo's gyroscope to overcome this problem.[15]

Tests conducted from 15 May to 8 July 1922 illustrated how radio control enhanced performance. While the aerial torpedo's human cargo sat and observed, another pilot in a nearby aircraft directed the weapon via radio signals. By providing occasional corrective signals, the pilot in the control aircraft guided the weapon to targets up to ninety miles away and then shepherded it back to the departure field, where the observer pilot took control and landed.[16] This process established a new role for airmen. The human component in the strike aircraft was functionally replaced by an exogenous agent who directed the bomber to its target. In actual combat, the cockpit would be filled with high explosives instead of a passive observer, and airmen would wage war from a remote-control console rather than the cockpit. A derivative of this phenomenon, remote-controlled unmanned drones, would appear again in the skies over the Third Reich and flourish in the skies over modern Iraq and Afghanistan.

In 1924 the Air Service's chief of plans was informed that radio control rendered the aerial torpedo "as a foremost machine of War."[17] It had matured into a true precision-guided cruise missile. Improving the range, accuracy, and flexibility of the aerial torpedo opened up new combat possibilities for airpower enthusiasts. The Air Service could now use unmanned weapons to interdict targets well beyond artillery range, such as ammunition dumps, staging areas, and, perhaps, enemy ships. The aerial torpedo piqued the interest of Major General Mason Patrick, the chief of the Air Service, and his deputy, Brigadier General William "Billy" Mitchell.

General Mitchell suggested in 1923 that aerial torpedoes should be employed against battleships, and General Patrick encouraged engineers at McCook Field to supply aerial torpedoes for the next round of aircraft-versus-battleship tests in August 1923.[18] The interest in sinking battleships with unmanned devices suggests that the Air Service was concerned with vindicating the concept of airpower rather than polishing the image of pilots. One imagines that General Mitchell sensed the delicious irony of sinking a warship with an Army aircraft dubbed the "aerial torpedo." He encouraged the Engineering Division to expedite progress on this weapon, yet the aerial torpedoes were unavailable for the 1923 antibattleship test due to ongoing modifications at the Sperry Gyroscope Company.[19]

By the mid-1920s the aerial torpedo had evolved from its early Kettering Bug phase to a larger, remote-controlled aircraft. Still, its inherently unstable aerodynamic characteristics and gyroscope limitations required provision for a safety pilot. The next phase of aerial torpedo development took place in the late 1920s and early 1930s, and it emphasized greater aerodynamic stability and alternative means of guidance. A 1927 appraisal of the aerial torpedo recommended the use of a more aerodynamically stable airframe, and the Air Corps looked to commercial aircraft since their design ensured maximum stability.[20] Using commercial planes as a source for aerial torpedoes also provided the benefit of a ready supply of these weapons—a sort of aerial fleet-in-being—during time of war. To test this concept, the Materiel Division purchased a Curtiss Robin monoplane and converted it into an aerial torpedo that relied on a simple azimuth gyroscope.[21] It had room for a safety pilot and could be loaded to the brim with explosives.

This phase of aerial torpedo evolution also explored new possibilities for remote control. Instead of control by a pilot in a nearby aircraft, the Air Corps sought ways to control the aerial torpedo from the ground. This displaced pilots even further from the air weapon by replacing their inputs with the expertise of ground-based radio technicians. In 1930 and 1931, engineers worked to control the aerial torpedo via radio beacons, just as Lieutenant Jimmy Doolittle had relied on radio beacons to guide him to the runway for the first blind landing in 1929. The Air Corps' goal was to fly the weapon along a radio beam toward the target. Another radio beam would intersect this flight path over the target. Once receiving this other signal, the aerial torpedo could either dive into the target area or drop its explosives and return along the path of the original radio beam. This tactic eliminated the need for clear visibility, leading the Air Corps to conclude that it could strike targets "the size of ammunition dumps and railroad yards at distances up to 200 or 300 miles" in fair and foul weather.[22] This new phase of the aerial torpedo, however, failed to achieve these operational goals before the program stagnated in 1932.

The aerial torpedo did not become a "foremost machine of War" in the Air Corps' interwar arsenal because of organizational difficulties and insufficient leadership that blurred focus on this weapon. The Materiel Division at Wright Field, the locus of research and development in the Air Corps, complained to the secretary of war for air in 1927 that the aerial torpedo project needed more organizational emphasis and "has not been pushed to the extent which it might have been."[23] No organizations were established to operationalize the aerial torpedo beyond the test stage or develop suitable doctrine to exploit its potential.[24] The project also lacked proper staffing, especially engineers and scientists. The aerial torpedo's development, furthermore, shared funding and manpower with related projects in instruments and navigation technology. Essentially, the Air Corps' research and development work was scattered in broad directions. The aerial torpedo program's efforts to couple radio control with an azimuth gyroscope, for example, were steered toward improving the navigation of manned aircraft in limited-visibility conditions.[25] The aerial torpedo, furthermore, received no funding in 1931 and 1932, so assets were funneled into the program from other projects related to instrumentation and automation.[26] In May 1932, with money tight throughout the military, the Air Corps mothballed the aerial torpedo for want of funds, convinced that the project would require years of expensive development.[27] Continued improvements in instrument technology and radio guidance contributed to its revival within a few years.

Revival

Developments in blind-flight technology in the early 1930s included enhancements to gyroscopes and autopilots as well as improvements in radio guidance. The new paradigm of instrument-based flight also elevated the status of engineers and required airmen to submit to artificial means of interpreting the environment and controlling the aircraft. This work contributed to the revival of the aerial torpedo. In May 1935, two years before Captain George Holloman's fully automated, hands-off flight and landing, Major General Benjamin Foulois, one of the Army's first pilots and now the chief of the Air Corps, argued that recent progress in automation technology "indicated that radio control of airplanes is not impracticable."[28] General Foulois directed the Air Corps Board, a special staff that researched various airpower-related issues, to determine the feasibility of developing the aerial torpedo.[29] The board concluded in October 1935 that recent technical advances in automation and radio control warranted resumption of the aerial-torpedo project. Although the chief of the Materiel Division questioned the high cost and practicality of the endeavor, the board maintained that this research promised "knowledge far more valuable than its cost."[30]

The Air Corps Board recommended purchase of commercial aircraft for adaptation as aerial torpedoes or drones for antiaircraft gunnery. The new chief of the Air Corps approved this recommendation but only in principle. Major General Oscar Westover, who replaced Major General Foulois in late 1935, encouraged experiments in remote-controlled flight but blocked the purchase of commercial aircraft for this purpose.[31] For the next two years the Air Corps used its own aircraft to conduct research in remote-control technology that enabled fully automated flight and benefited all-weather navigation as well as aerial torpedo development. In 1938 General Westover granted special priority to development of a small, low-cost aerial torpedo that could fly for twenty to thirty miles and be directed to its target by radio control.[32] He also authorized a design competition to allow any manufacturer to meet the requirements for this new weapon.[33] General Westover died in an airplane accident a few weeks after setting this policy, and his successor, Major General Henry "Hap" Arnold, inherited the program.

General Arnold came to the job fully aware of the aerial torpedo's twenty-year history. His interest in the initial efforts toward radio control had kept him in touch with the commander of the Air Service Engineering School in the early 1920s.[34] In 1930 he composed a lengthy report detailing the evolution of the aerial torpedo since the Great War, and he complained that the program stagnated in the early 1930s due to a lack of direction and leadership during research and development.[35] Now it was his turn to shape the future of unmanned aircraft. Initially, he kept with his predecessor's guidelines regarding the aerial torpedo's tactical role: it should carry 200 pounds of explosives for twenty miles, hit a target two miles square, and be designed to ensure high quantities at low cost.[36] The Army's adjutant general, however, challenged the Air Corps' specifications, noting that the "torpedo aircraft" had the same range as long-range artillery and contributed little military value.[37] Its limited capabilities made it little more than an extravagantly expensive, elaborately complex, and comparatively inaccurate artillery round.

General Arnold's rebuttal to the adjutant general revealed his belief in the potential of unmanned bombers and his faith in the ability of science and technology to continuously improve the air weapon. He assured the Army that modern manufacturers such as the Bell Aircraft Corporation and the Sperry Gyroscope Company possessed the expertise to exploit current technology and achieve progressively greater range and accuracy for the aerial torpedo.[38] Limiting initial requirements for range and accuracy provided a starting point for a program that could be expanded in the future. Arnold's foot-in-the-door strategy worked.[39] On 31 October 1938 the adjutant general approved funds for a design competition to meet the initial military specifications.[40] A little over a year later Arnold prescribed new specifications for the aerial torpedo: a range of 100 miles and an accuracy of a one-half mile diameter circle. Arnold also noted that payload capability and range might increase in future versions.[41] These military specifications could now be compared to a bomber rather than an artillery round.

The 1939 correspondence between General Arnold and Charles Kettering, inventor of the "Bug" during the Great War, provides further insight into Arnold's vision for an unmanned strike aircraft. "I am convinced we must continue the development of this weapon," he wrote Kettering, and it must be "suitable for controlling from another airplane in the air and from the ground."[42] Despite General Arnold's enthusiasm for an unmanned bomber and his plan to steadily increase its range and accuracy, progress in the aerial torpedo's development moved slowly. The Vega Airplane Company was the only serious respondent to the Air Corps' design competition in 1939, and the Materiel Command determined Vega's design held little promise. Kettering worked in the Research Laboratories Division of the General Motors Corporation during this time, and the Air Corps turned once again to his expertise. It signed a contract with General Motors on 14 February 1941 to develop ten aerial torpedoes for initial tests.[43]

In April 1941 use of the term aerial torpedo was banned in Army aviation. A joint Army–Navy conference reserved this label for devices dropped from aircraft into the water.[44] This was an early volley in the interservice rivalry over control of unmanned aerial weapons. The Air Corps selected the unwieldy name "controllable bomb, power driven."[45] The General Motors version of the aerial torpedo, therefore, was called the "controllable bomb, power driven, General Motors, type A-1," or "GMA-1." A small, high-wing monoplane, the GMA-1 bore a marked resemblance to Kettering's Bug of World War One. Unlike the original Bug, it contained a gyroscope-based autopilot integrated with a radio-control mechanism and boasted a range of 200 miles with a 500-pound payload.[46] Tests at Muroc Lake, California, in 1941 and 1942 established that this version of an unmanned bomber still required significant improvements. Despite the use of an autopilot and radio control, the first several units crashed shortly after takeoff. The GMA-1 showed more promise, however, as its human operators gained proficiency in a new method of flight that integrated automatic controls with remote radio control.

In one test flight of nearly two hours, the GMA-1 encountered turbulent air that tumbled its gyros and sent it rolling out of control. Technicians swiftly regained control by switching to manual, radio-based inputs. This illustrated a drawback of fully automated aircraft since gyroscope-based feedback systems could maintain control only in stable conditions.[47] It also highlighted the human component's role in this new manner of flight. Pilot input was still required somewhere in the cybernetic control loop to provide corrections in dynamic, unstable conditions. A colonel overseeing the project informed General Arnold that operational employment of the GMA-1 would require large numbers of "pilot-engineer operators" and recommended that pilots assigned to normal bombardment groups undergo additional training to operate this unmanned weapon.[48] This new type of flying skill would also require proficiency in television-based control. Equipped with a television camera in its nose, the GMA-1 could be guided directly to its target by operators on the ground using a television monitor.[49] The controllable bomb, power driven became an expensive, precision-guided weapon requiring intensive human oversight. The Army Air Forces chose not to field it in combat operations.

Several factors weighed against further development of the GMA-1. Its limited range left many strategic targets out of reach, and a 500-pound payload was effective only when delivered with pinpoint accuracy. Precision, however, required human operators in nearby aircraft and a heavy, complicated television apparatus, as well as favorable weather. Development of the weapon and training in how to use it entailed significant cost as well. The Army Air Forces thus terminated the GMA-1 contract in May 1943 in favor of other prototypes of unmanned bombers. At the end of 1942, for example, General Arnold approved research into "interstate power-driven bombs of 2,000 lb. type [payload] with range of at least 1,000 miles."[50] This suggested that even fairly early in the war he envisioned the strategic use of unmanned bombers. Similarly, in early 1943 the Army Air Forces sought development of unmanned weapons with explosive loads of 4,000 pounds and ranges up to 1,500 miles.[51] The push for such weapons early in the war revealed a significant interest in the alternative of unmanned aircraft. This interest manifested itself in different ways throughout the rest of the war and helped establish the paradigm of unmanned flight in military aviation.

Robot Bombs: The Weary Willies Go to War

On 10 August 1944, Joseph Kennedy Jr. wrote from England to his younger brother, John: "Every day I think will be my last one here, and still we go on. I am really fed up but the work is quite interesting. The nature of it is secret. . . . Tell the family not to get excited about my staying over here. I am not repeat not contemplating marriage nor intending to risk my fine neck . . . in any crazy venture."[52] In reality, Lieutenant Kennedy *was* preparing to risk his neck in a crazy venture. Two days after writing his letter, he and a crewmate climbed into a B-24 Liberator bomber packed with high explosives. Their top-secret mission: take off, raise the landing gear, climb to a safe altitude, adjust the engines, engage the autopilot, arm the detonator for the massive 21,700-pound payload, and parachute to safety before crossing the English Channel. The unmanned bomber would then be flown under remote control by another pilot in a nearby aircraft and steered into a German V-2 missile launch site in northwestern France. But something went wrong. As Kennedy and his crewmate were arming the flying bomb and preparing to jump out, observers witnessed a tremendous explosion that obliterated the aircraft. The bodies were never found.[53]

The mission of flying a heavy bomber via remote control directly into a target refashioned the role of the pilot in aerial bombardment. This change in the pilot's portfolio was not a sudden development since the Army's first pilotless bomber or cruise missile originated in 1918 as Kettering's Bug, and during the 1920s various forms of the so-called aerial torpedo placed unique demands on pilots. Safety pilots monitored and adjusted automation devices to mitigate development costs, and pilots at ground stations or in "mother ship" aircraft learned the art of remote control. In 1935 the Sperry Gyroscope Company suggested that pilots could fly aerial torpedoes to a desired altitude, set the automatic controls, and bail out.[54] In late 1941, the Army Air Forces Experimental Engineering Section recommended that any aircraft could be converted into a "power driven controllable bomb" and operated much like a suicide bomber, only without the inconvenient suicide.[55] By the end of 1943, the Army Air Forces applied its experience with unmanned flying bombs, remote control, and television guidance to a new plan: the conversion of war weary, obsolete aircraft into unmanned "robot bombs."[56]

The Special Weapons Branch of the Army Air Forces Materiel Division worked to develop a remote-control apparatus for the B-17 Flying Fortress and B-24 Liberator bombers in early 1944.[57] Codenamed "Castor," this project seemed a plausible use for battle-worn aircraft that still possessed an operative automatic flight control system but had various forms of damage or structural fatigue that required expensive repair or replacement. Engineers developed a dozen remote-control units suitable for the B-17 and B-24, and they also modified a television camera for installation in the nose of the aircraft.[58] By peering at a crude television monitor and manipulating remote controls, airmen could theoretically steer Castor bombs into important targets. Shortly after Allied forces pushed into Normandy in June 1944, General Spaatz, one of the most ardent proponents of strategic bombing and the commanding general of the US Strategic Air Forces in Europe (USSTAF), encouraged the expeditious development and employment of Castor. He had a special purpose in mind: destroy German V-1 and V-2 launch sites in the Pas de Calais region. Previous attempts with manned bombers had failed to crack open these hardened targets and stem the barrage of V-1's against London. General Spaatz labeled this new approach Project Aphrodite.[59]

Aphrodite was a joint venture with the US Navy. The Navy had experimented with radio control and television guidance of drone aircraft during the late 1930s and World War Two. This included the Navy's Special Task Air Group One which deployed television- and radio-guided "torpedo drones" in the South Pacific in 1944.[60] Although the Navy limited this work to smaller aircraft, in

July 1944 it sent a PB4Y, the naval version of the B-24 Liberator heavy bomber, to participate in General Spaatz's campaign against the V-1 and V-2 sites.[61] Lieutenant Joseph Kennedy Jr. volunteered for this top-secret program and joined a small Navy contingent that worked alongside Army Air Forces personnel to employ remote-controlled bombers. A series of missions occurred in the months after Lieutenant Kennedy's tragic death: a second B-24 was launched in August and ten Aphrodite B-17s were expended between September and December. During this period Aphrodite became known as the "Weary Willie" or "Willie" program. The unmanned B-17 was dubbed "Baby," and its control aircraft "Mother." Although flying a Baby into targets via remote control appeared simple in concept, various factors such as human error, equipment failure, poor weather, and enemy defenses confounded this task.

Two Weary Willie missions launched on 11 September 1944 experienced different problems. The remote-control apparatus malfunctioned in one Willie Baby, plunging it into the English countryside before the pilot could bail out.[62] In the other sortie, the pilot had difficulty stabilizing the Baby's altitude due to a faulty pitch sensor on the automatic flight control equipment. Unable to correct the flying bomb's descent, he bailed out just over 1,000 feet above the ground. His parachute opened with little altitude to spare and nearly caught in some power lines. The Baby leveled off to a stable altitude of just 300 feet and flew toward its destination across the English Channel. The Mother aircraft, however, had trouble in maneuvering the Baby into the target: they missed on the first pass due to the defective pitch control and circled around for another attempt, allowing time for a German gun crew to zero in on the low-flying airplane. The flak gunners hit their mark at near point-blank range but lost their lives due to the proximity of the resulting blast.[63] Other difficulties emerged in later missions.

The remote-control equipment worked properly on the 14 September 1944 missions of two Willie robots, but human error rendered both missions ineffective. One Mother shepherded its Baby over 400 miles, but the television operator failed to tune the receiver correctly. The Willie Mother circled over 3 miles above the Willie Baby with only a fuzzy image of the target. The human operator missed the target by 800 feet.[64] In the other mission, the operators misjudged the difference in speed between the high-flying Mother and the low-flying Baby and lost sight of it somewhere over the French coast. The Mother aircraft circled back but could not find its errant Baby. The television image indicated it was somewhere over the sea; still unable to locate it, the controllers terminated the mission by flying it into the water.[65] These missions demonstrated that effective operation of this complex technological system required the human component to develop new skills and adapt to a significantly different type of military flight.

With proper operation of all the system's human and machine components, combined with minimal interference from the enemy, Willie did achieve some successes. General Spaatz sent his new weapon against the heavily fortified submarine pens at Heligoland on 15 October 1944, and Baby found its target for the first time. The B-17 hurtled into a complex of buildings, shattering two and one-half acres of structures and damaging other buildings within several hundred feet.[66] This mission helped justify some of the optimism that had been invested in the Weary Willie program. Willie enjoyed a "1A" priority in the fall of 1944, and Lieutenant General Jimmy Doolittle, commander of the Eighth Air Force, informed General Spaatz in early November that ten B-17s and fifteen B-24s would come available every month for conversion to cruise missiles.[67] The air force further authorized procurement of 550 control units for future use.[68] Despite these enthusiastic efforts, the new application of the air weapon encountered the same nemesis that thwarted manned bombers: poor weather.

The Continent's notoriously bad weather was a powerful ally to Germany's air defenses. Inclement weather obscured targets, grounded fleets of manned bombers, and hindered the Weary Willies as well. On 31 October, for example, two Willie Babies were guided to Heligoland for another attack, but clouds blanketed the target area. One Baby disappeared in the weather and crashed into the North Sea. The other Baby, in apparent contradiction to Army Air Forces rhetoric about the importance of precision bombing, was "dispatched on a compass heading for Berlin" and never seen again.[69] On 5 December the weather intervened again to cause significant problems. Two Babies were sent into Germany, but their Mothers could not find the designated targets due to cloud cover. One Baby was sent through a small break in the clouds to raze the town of Haldorf, but it hit short of its target and caused little damage.[70] Operators surmised that the other Baby experienced some weather-related carburetor icing. This caused the engines to lose power, and the bomb-laden Flying Fortress executed a surprisingly gentle descent into a ploughed field and skidded on its belly to a stop. Fully intact, it was an early Christmas gift to Wehrmacht intelligence officers and a deep embarrassment to the Army Air Forces. After waiting for a break in the weather, multiple fighter and bomber missions were sent to find and destroy the lost Baby. The hunt continued for several days, but the Baby was never found.[71]

Castor technicians at the Army Air Forces Materiel Division sought means to overcome weather-related difficulties. One method involved rigging Baby aircraft with a television camera and transmitter in the cockpit.[72] This camera sent a pilot's-eye view of the flight instruments to a receiver display in the Mother aircraft. Therefore, the Mother's pilot flew his aircraft on instruments, and the Baby's pilot sat at a console in the Mother and flew the Baby aircraft on instruments via a crude television image. The Baby's autopilot reduced some of the workload of its exogenous remote-control pilot, but this still presented a new challenge in the human control of the air weapon. Such flight in inclement weather involved an extravagantly complicated feedback loop: a human responded to a televised image of cockpit instruments that represented a distant machine's relationship to the Earth and transmitted his inputs to an artificial mechanism that moved the controls and altered the readout of the instruments, which then provided a new image to the human mechanism. This complex cybernetic affair was a new form of flight that anticipated the actions of Predator and Reaper drone pilots in the modern Air Force. It physically displaced the airman from the aircraft and functionally replaced his senses with sensors, his judgment with gyroscopes, and his touch with telemetry.

An additional development in unmanned bombers appeared in conjunction with the Weary Willie program: Willie Orphan. Willie Babies were steered about by Mother aircraft, but Orphans had no Mother and depended on radar stations, radio beacons, and controllers on the ground for guidance to their target. When radar operators deemed the Orphan to be somewhere in the vicinity of its target, they would simply fly it into the ground and hope for the best. In late September 1944, Brigadier General Grandison Gardner, the head of the Proving Ground Command, recommended to General Arnold that Willie aircraft be sent to enemy targets using radar control. "The Proving Ground is poor in scientists and engineers and we have blundered along pretty slowly trying it out," he claimed, "[but] by taking off with pilot aboard we can control these airplanes for as much as fifty miles."[73] This put a variety of German towns in range from radar stations in northern France, and General Gardner noted that such weapons would exploit winter weather. Except for the takeoff, Orphans could operate in zero visibility and enjoy relative immunity from German fighters when cloaked in fog or clouds. General Arnold sent fifty radar guidance kits to Europe and instructed General Spaatz to consider the project.[74]

General Spaatz, who would eventually ascend to General Arnold's position in 1946, appeared willing to test Willie Orphan against "fortified German cities."[75] The fact that at least two Willie Babies had been steered to population centers such as Haldorf and Berlin suggested that the Army Air Forces was not entirely dedicated to its proclaimed doctrine of precision bombing against specific military and industrial targets. Even with ground-based radar guidance Orphan lacked precision, and launching it toward large, civilian-populated targets in Germany paralleled the Wehrmacht's lobbing of V-1s and V-2s into southern England and the Low Countries. General Spaatz, however, balked at employing Orphan as a terror weapon.[76] He wanted it to hit specific targets to the maximum extent possible, and he inquired about how radar-guided bombers could be steered into their targets once reaching the desired area.[77] The Ninth Air Force, an organization within General Spaatz's command, claimed that Weary Willies lacked sufficient accuracy unless steered from Mother aircraft; moreover, it argued that fielding a large number of Orphans would siphon resources from the ongoing manned bomber offensive against the Reich.[78]

The Army Air Forces shipped seventy-five remote-control kits for Weary Willies to the European Theater of Operations, yet they remained unpacked at the end of the war.[79] Although General Spaatz acknowledged the potential value of the weapon system, several factors hobbled its practical employment as a strategic weapon. Orphans and Babies could not venture above 10,000 feet since ice tended to clog the oil regulators in the engines unless a human was onboard to make constant adjustments.[80] Flying below 10,000 feet limited their potential range and made the robotic bombers low, fat targets for a potent German air defense. Even if a slow-flying Orphan or Baby reached its target area, it still faced withering fire in the final approach to the ground. Moreover, strategic use of the weapon required long-range flights that exceeded Orphan's radius of operation and demanded ideal weather conditions for Baby aircraft.[81] Overall, war-weary unmanned bombers offered the promise of high-tech warfare but failed to deliver in terms of consistency, flexibility, and precision. Their introduction into actual warfare, however, promoted technological development that displaced the human component and focused on the unmanned projection of power. While the Army Air Forces experimented with its Babies and Orphans, it also developed other members in the family of unmanned weapons.

Jet Bombs

Airmen described unmanned aircraft in numerous, interchangeable terms during the interwar years and World War Two, including aerial torpedoes, pilotless bombers, pilotless aircraft, cruise missiles, robot bombs, and controllable bombs, power driven. During World War Two, the Army Air Forces Board defined unmanned devices subject to control after launch as "guided missiles." Guided missiles could, among other things, "complement, supplement, or supplant long range strategic bombardment."[82] The notion that such weapons could replace manned bombers suggests a belief that airpower was not confined to traditional methods of piloted flight. Aircraft, manned or unmanned, were devices that enabled warfare through the atmosphere and served as a means to an end. In this instrumentalist view, the human component could be usurped by technology that was more effective or efficient. This concept had been reified in various forms since the First World War: the Bug, the Aerial Torpedo, the General Motors A-1, Weary Willies, and the "Jet Bomb 2," or JB-2.

The Germans, not the Americans, first introduced a pilotless bomber in combat. Launched into southern England from sites in northwestern France, the Vergeltungswaffe-1 (Vengeance Weapon 1, or V-1) was an unmanned, unguided, gyro-stabilized aircraft propelled by a jet engine Churchill described as a "new and ingenious design."[83] The Germans sent the first V-1s toting a one-ton payload against London on 13 June 1944. The flying bomb came to be known by its engine's distinctively loud, sputtering noise and earned the nickname of "buzz bomb" or "doodlebug."[84] Their indiscriminate torrent claimed the lives of over 6,000 Britons during the next few months.[85] The V-1 precipitated the development of a copycat version by the Army Air Forces: the "Jet Bomb 2" (JB-2). Using salvaged parts from V-1 wreckage, the Army Air Forces reverse-engineered the German design and instituted a crash program to operationalize the JB-2 as soon as possible. This included construction of a manned version to expedite the development and refinement of the cruise missile's flight characteristics.[86] Two months after the first V-1 buzz bomb jetted into London, the Materiel Command ordered production of one thousand JB-2s and encouraged the development of the operational doctrine and organizational capacity to employ it in combat.[87]

Although the JB-2's rapid development was driven by the appearance of the V-1, its actual integration into combat operations met with resistance. In

July 1944, General Spaatz thought that the achievement of air superiority over Germany and the rapidly changing nature of the ground war limited the need for employment of the JB-2. He noted it might come in handy if German defenses forced a stalemate; in that case, JB-2s could be launched into the industrial districts of the Ruhr.[88] Creating legions of cruise missiles, however, would also impose significant logistical penalties. The USSTAF in Europe worried that mass employment of JB-2s would strain manpower and equipment supplies, especially in terms of maintaining adequate levels of ammunition for artillery and general-purpose bombs.[89] The JB-2 thus competed for limited resources with other weapon systems already employed in the European Theater of Operations.

Despite the USSTAF's concerns over assigning resources to the JB-2, by December 1944 the Army Air Forces contracted for production of 2,000 units. In January 1945, General Arnold gave the program even greater emphasis. He ordered 100 launches per day by September 1945, 200 per day by October, and a blitz of 500 per day by January 1946. To meet these demands, General Arnold authorized procurement of 75,000 JB-2s, nearly four times the total number of buzz bombs launched by Germany against Britain and targets on the Continent.[90] Although the chief of the Army Air Forces emphasized development of a remote-control capability similar to the technology used for Willie Orphan, he stipulated that work on remote guidance should not delay the launch of unguided versions of the JB-2.[91] The decision to emulate German tactics by dispatching unguided, unpiloted weapons to indiscriminately bomb the enemy indicates that Army Air Forces leadership was not bound by its own rhetoric concerning the virtues of precision strategic bombing, nor was it constrained by the notion that victory must come at the hands of pilots. For these airpower enthusiasts, pragmatism trumped principle.

In August 1944, the assistant chief of the Air Staff predicted that the new type of weapons known as guided missiles could be of "decisive importance." Similarly, a few months later the USSTAF acknowledged that the JB-2 might be a "terror weapon when used on a small scale," but it also argued that the JB-2 "might be decisive" if launched in massive numbers.[92] In a February 1945 letter to Vannevar Bush, President Roosevelt's influential director of the Office of Scientific Research and Development, the deputy commander of the Army Air Forces, Lieutenant General Barney Giles, contended that Germany could have defeated England by launching a thousand or more V-1s per day. This view suggested a belief that even imprecise, indiscriminate bombardment could serve as an effective strategy. It also revealed his faith in the JB-2 as part of "a new family of very long range weapons whose capabilities will profoundly affect future warfare, and especially aerial warfare."[93] Aerial warfare, in his view, was no longer the sole regime of piloted aircraft, nor did it require adherence to the doctrine of precision strategic bombardment. General Giles also observed that Japan presented an array of acceptable targets for waves of JB-2s, guided or unguided. He anticipated 1,000 launches per month by the end of 1945, with round-the-clock JB-2 strikes from ships and nearby islands supplementing other operations against the Japanese.[94]

The momentum toward employment of the JB-2 faded as the Allies gained strategic advantage in Europe in 1944 and 1945. Deliveries began to trickle into the European theater of operations in January 1945, but the Army Air Forces canceled production on V-J Day and closed the program on 1 March 1946.[95] Although the JB-2 never gained operational experience, one of the program's key goals was, as the Air Staff put it in August 1944, to "gain experience for future reference."[96] With an eye on future independence from the Army, air force leaders sought new methods to command the air, even methods that displaced the human from the machine. These methods demonstrated that the air force's key mission of strategic bombardment and the ambitions of airpower enthusiasts were no longer tied to the paradigm of manned flight.

Pilotless Weapons: An Institutional Imperative

A primary goal of airpower enthusiasts within the Army in the 1930s and 1940s was to gain independence and establish a separate, autonomous air force. This required validating the usefulness of airpower. The Army Air Forces thus employed the air weapon to destroy objectives in the battle area, interdict enemy activity behind the battle lines, and, most importantly, bomb strategic targets deep in enemy territory.[97] For advocates of an independent air force, strategic bombardment was the essence of their art. They presumed that technologically advanced bombers would cripple the enemy's will and ability to fight. In the spirit of Air Corps icon General William "Billy" Mitchell, they believed airpower could "not only dominate the land but the sea as well."[98] Technological advances in automation, remote control, and cruise missiles revealed that airpower could include unmanned weapons. These alternative forms of airpower strayed from the tradition of piloted aircraft, yet they offered another means to achieve the air force's objectives. To help secure independent status, therefore, the air force sought dominion over pilotless weapons.

In 1967 Theodore von Kármán, friend and scientific adviser to General Arnold during World War Two, remarked, "We used the term 'pilotless aircraft' to cover all types of missiles, so as to prevent the project from falling into the hands of the Army, from which the Air Force was about to separate."[99] Many in the Army's ground forces considered surface-launched missiles as another form of long-range artillery and considered the air force's interest in aerial torpedoes and missiles as a trespass on the soldier's turf. As noted earlier, in 1938 the Army's adjutant general considered the range-limited aerial torpedo an expensive gadget that achieved the same effect as traditional cannon. Similarly, the development of cruise missiles such as the JB-2 overlapped a proposed Army Ordnance project on long-range missiles. The Army Ground Forces sought development of "major caliber guided missiles" that delivered a 2,000-pound warhead deep into enemy territory, and in August 1944 the assistant secretary of war for air advised the Army Air Forces to strengthen its claim to such technology or risk losing it to the Army Ground Forces.[100]

One method the Army Air Forces used to assert sovereignty over guided missiles was to point out the similarities and functional connections between missiles and aircraft. Dr. Edward Bowles, "Expert Consultant to the Secretary of War," noted that the Army Ordnance's project involved a *wingless* missile, whereas the air force's version resembled an aircraft.[101] Both served a similar purpose, yet their different construction reflected the institutions that produced them. The soldier's missile was a cylinder with fins; the airman's missile was an airframe with wings. For the Army Air Forces, such differences validated its continued development of "pilotless aircraft type projectiles."[102] When the Army Air Forces got wind of the Army Ground Forces' interest in ordering 25,000 JB-2s, it cited several reasons why this unmanned weapon belonged in the hands of airmen instead. The JB-2 was kith and kin to aircraft since it had wings for aerodynamic lift, used methods of control similar to other air force equipment such as Weary Willie, and enabled attack of strategic targets such as industrial areas or population centers. Since strategic attack was the central mission of airpower, the Air Staff argued, this weapon should naturally come under air force control.[103]

The Army attempted to settle the brewing quarrel between its Air Forces and Ground Forces over the development of guided missiles. In October 1944 Lieutenant General Joseph T. McNarney, the deputy chief of staff of the Army, assigned the Air Forces responsibility for the research and development of missiles deployed from aircraft and missiles launched from the ground that relied on "the lift of aerodynamic forces." Surface-launched missiles that "depend for sustenance primarily on momentum" were assigned to the Ground Forces.[104] McNarney's directive proved to be a shaky compromise and led to further disputes. The Ground Forces challenged the proscription on aerodynamic lift when developing its Hermes missile. German V-2 ballistic missiles had been falling at supersonic speed on Antwerp, London, and other areas since September 1944, and in November the Army Ordnance Department initiated its own version of the V-2 called Project Hermes.[105] When one Hermes scientist suggested the addition of wings, the Air Forces cried foul.[106] Wings, after all, were the construction and jurisdiction of airmen. Yet airmen also disputed the McNarney protocols.

Shortly after the end of the war, with interest in missiles intensifying in the Army's air and ground components, General Spaatz challenged the McNarney directive by recommending that Army Ordnance's jurisdiction extend only to missiles that, like artillery rounds, contained no method of guidance during flight.[107] A month later the Army Air Forces deemed General McNarney's compromise invalid and set its own policy: "It is now apparent that all missiles requiring guidance in flight, whether sustained by the lift of aerodynamic forces or by the momentum of the missile, will be pilotless aircraft operating on principles of aerodynamic control and stabilization."[108] The air force's imperative to control various forms of missiles thus helped precipitate a change in the definition of flight. Flight was not just the transit of a manned aircraft through the air. Rather, it also included unmanned devices controlled by aerodynamic or stabilizing forces and subject to some form of guidance. Such a definition of flight and the labeling of missiles as "pilotless aircraft" allowed airpower enthusiasts to claim authority over a new family of weapons and enhance the air force's institutional power.

In addition to emphasizing that missile development should belong to the established experts on flight, the Army Air Forces sought to assert control over unpiloted weapons by rushing them from the drawing board to the combat theater. Accordingly, the Air Staff recommended using the JB-2 cruise missile in combat operations as soon as practical.[109] In the calculus of the staff, the first institution to develop or employ a device gained an advantage for future claims of priority regarding such technology. This notion also materialized with regard to ground-launched missiles that targeted enemy aircraft as well as terrestrial targets. Major General H. R. Oldfield, the Army Air Forces special assistant for antiaircraft, urged the expedited development of ground-to-air missiles by the air force.[110] By doing so, the Army Air Forces could steal a march on the Army Ground Forces and boost its claim for control over the Army's antiaircraft artillery. Accelerating the development of unmanned weapons would help cement the air force's claims to various technologies of pilotless power projection, and this gambit would also benefit from organizational changes in the Army Air Forces.

By the summer of 1944, the Air Staff recognized the need to exploit the potential of unmanned weapons and worked to establish an organization dedicated to their development.[111] The Army Air Forces formally established the Pilotless Aircraft Branch in February 1945 and directed it to coordinate with military and civilian agencies and integrate new ideas and technologies in the trajectory toward unmanned weapons.[112] The Pilotless Aircraft Branch also provided parity with the Navy. During the war, the Navy poured considerable effort into guided missile technology at its Ordnance Testing Station. Airmen envied this amply funded, well-organized agency, and one general remarked, "Our dilly-dallying methods have, in fact, thrown many projects . . . into the Navy Department with consequent later procurement difficulties on our part."[113] The Pilotless Aircraft Branch helped bring diverse methods into sharper focus and provided an organizational home for the transition to a new type of warfare. Its very existence also demonstrated a central irony familiar to drone advocates today: the best interests of a pilot-led institution were served by the development of unpiloted machines.

Conclusion

In the same press conference of 18 August 1945 where he mentioned that an airman seated at a control console could fly a television-guided bomb fifteen miles away, the commander of the Army Air Forces touted the continuing development of long-range, jet-propelled guided missiles. "These Buck Rogers things I'm talking about," General Arnold proclaimed, "are not so Buck Rogerish as you might think."[114] Just a few days after V-J Day, the nation's top airman had his eyes on the future. He declared, "The past is a dead horse" and argued that a new era of weapons lay ahead: manned and unmanned supersonic aircraft, target-seeking warheads on guided missiles, and antimissile missiles.[115] Dead horse or not, and thanks in part to galloping strides in science and technology, the past had delivered General Arnold and the Army Air Forces to the point at which the definitions of flight and airpower now included a pilotless means of commanding the air.

General Arnold's forward-looking comments were made possible by advances in unmanned technology that stretched from World War One through World War Two. The first cruise missile came into infancy during the Great War, and this so-called aerial torpedo evolved in parallel with aviation's increasing reliance on instruments, autopilots, and navigational aids. This alternative form of power projection displaced pilots from their cockpits and shaped a new form of flight. Airmen involved in these projects learned how to "fly" aircraft from a remote-control console and developed methods to integrate radio commands with automation devices.

Although the trajectory toward greater automation opened the door to unmanned weapons, such weapons were not a predetermined or inevitable aspect of technological advances in airpower.[116] Instead, their development and use were shaped by multiple factors, such as specific political and military objectives and institutional interests. For example, the necessity to stop Nazi terror weapons through strikes on hardened V-1 and V-2 launch sites occasioned the employment of unmanned B-17 and B-24 bombers, and the effort to achieve technological parity with German missiles catalyzed development of the JB-2. Moreover, the air force redefined flight as both a manned and unmanned activity involving control of aerial devices in order to claim institutional jurisdiction over an expanding array of airborne weaponry.

The Army Air Forces' early effort to secure premier status as the nuclear strike force provides another example of how its quest for power and independence altered the airman's role. By the fourth anniversary of Pearl Harbor the Army Air Forces aimed to merge its "Pilotless Aircraft Program" with an entirely new technology: the Bomb. Engineers sought technical data from Major General Leslie Groves, the director of the Manhattan Project, for "the development of a ground-to-ground pilotless aircraft carrying an atomic warhead."[117] Ground technicians, not pilots, could deliver future atomic strikes. The effort to couple these advanced technologies affirmed airmen's belief that redefining man's role in flight could serve their military goals and institutional ambitions.[118]

The Modern Pilot, Redefined

A modern pilot is like the corpse at a funeral: his presence is traditionally expected, but he doesn't do much. Or so it may appear given the changes in what some pilots actually do in terms of physical control. Despite their commanding position behind a locked cockpit door or strapped in an ejection seat, most of the time today's pilots serve as observers and managers of complex cybernetic systems that do the actual flying.[1] Occasionally they grip controls, jockey throttles, and apply manual stick-and-rudder skills, yet even these actions are adjudicated by computer logic in fly-by-wire aircraft. Modern airline pilots rarely "hand-fly" the aircraft as they comply with company directives to surrender control to the autopilot except for takeoff and landing.[2] Sometimes even the landings are machine-flown.

Increasingly, pilots serve as information managers and type commands into a computer as they monitor displays connected to various onboard sensors linked with Global Positioning System (GPS) satellites. One Airbus A310 airline captain observed, "I had evolved from the hands-on flier of my earlier years to a systems manager, controlling the plane with a flight management keyboard," and he recalled the pilots' joke that you "had to be able to type 50 words a minute" to pass the Federal Aviation Administration's evaluation of your flying skills.[3] No doubt a growing number of US Air Force pilots are superb typists since more will be trained to control unmanned aircraft from ground-based computer consoles than will be trained to fly traditional (yet highly computerized and automated) bombers and fighters.[4] Pilots trained to fly manned aircraft, military or commercial, can succeed only when they commit to becoming managers, often hands-off managers, of a highly computerized and heavily automated system.

This chapter connects past and present. It illustrates how early efforts to overcome human limitations are reflected in modern civilian and military aviation. What do pilots really do nowadays at the leading edge of technological change, and what are their new expectations and challenges? By examining the pilot's role in modern commercial and military aviation, we can better understand the fundamental redefinition of flight and why the history and modern trajectory of aviation is a story of the pilot's transition from physical master to modern manager.

Commercial Aviation: The Problem with (Auto)Pilots

The *New York Times* declared in July 1933, "Commercial flying in the future will be automatic." This claim was based on famed aviator Wiley Post's heavy reliance on radio navigation and an automatic pilot during a record-breaking seven-day circumnavigation of the globe in his specially equipped *Winnie Mae*. A "robot pilot" was Post's lone companion; it let him "sit relaxed in his chair, an overstuffed one specially installed, without much regard for the stick technique" and "held the plane on a set course without aid from the pilot."[5] Post's use of a radio compass to fly great distances with considerable accuracy, combined with his extensive use of automatic control to counter fatigue, demonstrated the value of machine-based control. The *Times* described this as "a revelation of the new art of flying," and its prediction of a highly automated future of flight was largely correct.[6]

The *Winnie Mae*'s historic flight was a telling waypoint in the trajectory of increasingly sophisticated automatic control technology designed to overcome human limits. The array of subsequent advances in automation during the 1930s and World War Two improved the safety, performance, and potential of aircraft and created new roles for pilots as managers of increasingly complex systems. These advances were illustrated shortly after the war by the first transatlantic crossing of a fully automatic "robot-piloted plane." Twenty years after Charles Lindbergh's epic 1927 crossing in the *Spirit of St. Louis*, a Douglas C-54 Skymaster equipped with a "mechanical brain" took off from Newfoundland, flew 2,400 miles, and gently touched down at an airfield near London without a human operator ever touching the controls. The crew of nine sat and observed.[7] One passenger, Major Thomas Weldon from the Flight Test Division at the Air Force's Experimental Center at Wright Field, noted "this equipment has enormous potentialities for commercial and military aviation; it changes the pilot's role from one of operator to that of supervisor."[8]

The phenomenon of push-button, automated, human-supervised flight saturates modern civilian and military aviation. Although every airline pilot still knows how to manually fly the machine, hands-on control is generally avoided since no pilot can control an aircraft as smoothly or efficiently as an automatic flight control system. Accordingly, airline companies require pilots to engage the autopilot for most flight and essentially serve as hands-off safety monitors and system managers.[9] Pilots are expected to exert manual control if necessary during an emergency or unusual circumstance even though they rarely practice such control, and their actions may still be limited by automatic systems designed to prevent mishandling by humans. The pilot thus remains a vital component in the modern airliner but possesses a skillset that reflects a fundamental shift in the human's role. One way to understand the role and expectations for today's pilots is to examine a tragedy where pilots failed to safely interface with and manage the highly automated systems of a modern airliner.

Air France Flight 447 departed Rio de Janeiro for Paris on the evening of 1 June 2009, with 216 passengers and 12 aircrew. Three experienced pilots occupied the Airbus A330's cockpit. The captain, age fifty-eight, had logged nearly 11,000 total flying hours and 1,747 hours as an A330 captain. The copilot in the left seat at the time of the accident had long familiarity in the A330 (4,479 hours) and a total of 6,547 hours. The copilot in the right seat, approximately half of the captain's age at thirty-two, had accumulated 2,936 total hours with 807 hours in the A330.[10] The records of all three pilots confirm standard training in emergency procedures dealing with airspeed indicator malfunctions, stalls, and unusual aircraft attitudes. Nevertheless, their training and experience proved inadequate to maintain control of the Airbus when its automated control systems were disrupted.

The Airbus performed an uneventful departure and established a stable cruise altitude for its nighttime transatlantic journey. Its redundant Flight Management Guidance and Envelope Computers operated normally: they ensured the auto-throttle system maintained precise airspeed control and that the autopilot kept the aircraft on its programmed altitude of 35,000 feet on the exact course specified by the navigation computer.[11] A key part of this automated system, the Flight Director, offered a visual display of data from the Flight Management Guidance and Envelope Computers. This took the form of two bright green crossbars, one vertical and one horizontal, superimposed upon an artificial horizon in front of the pilot and copilot.[12] The Flight Director crossbars display what the computer determines to be the correct three-dimensional position of the aircraft relative to the surface of the Earth. If the vertical crossbar is left or right of center, it means the aircraft needs to turn left or right, respectively; if the horizontal crossbar is above or below the horizon, it means the aircraft's nose needs to pitch up or down. Under normal automated conditions, the Flight

Director crossbars are centered like crosshairs in a riflescope to indicate the aircraft is on its precise, planned flight path relative to the ground. This reliance on an instrument-based version of reality traces all the way back to the gyroscopically driven artificial horizon used by Lieutenant James Doolittle in his pioneering "all blind" flight in 1929, and it is a more sophisticated version of the early flight director technology of the 1930s which superimposed a small airplane silhouette on the instrument panel's artificial horizon.

Even when pilots hand-fly the Airbus, the Flight Director crossbars still tell the pilots where to point the nose of the plane. This technological augmentation is so embedded in cockpit operations that the Airbus operator's manual mandates the Flight Director (FD) to be turned on before takeoff and stipulates: "When the FD's are used, given the degree to which the A/THR [auto-thrust] modes depend on the vertical modes, the FD orders *must be followed*."[13] The manual acknowledges pilot agency, however, by stating, "An automatic system is, and must remain, an aid" and, "When the operation of the automatic systems does not correspond to the pilots' expectations and if the cause is not immediately analyzed without any ambiguity, the system(s) in question must be disconnected."[14] In the final minutes of Air France Flight 447 the pilots were apparently unable to distinguish between automatic indications and reality. Confusing and intermittent guidance from the Flight Director compounded their crisis.

The pilots became mildly concerned about some persistent light turbulence as they monitored the aircraft's progress on its northeast heading through a layer of clouds.[15] The captain, a veteran of Atlantic crossings, remained confident about the situation and decided to begin an authorized rest period. He put the copilots in charge and left the cockpit for a nap. Minutes later the Airbus encountered meteorological conditions that clogged its airspeed sensors, called pitot tubes, with ice. The pilots were unaware their Airbus could no longer sense its speed through the air. This confounded the computerized flight control system. Per their programming, the Flight Management Guidance and Envelope Computers disconnected the autopilot since the airspeed data was unreliable. The sudden disengagement of the autopilot ignited a variety of alerts in the Airbus cockpit: the "Master Warning" light on the Flight Control Unit illuminated, an "AUTO FLT AP OFF" warning appeared on the cockpit's Electronic Centralized Aircraft Monitoring display on the instrument panel, and a "cavalry charge" alarm blared through the cockpit speakers.[16] The copilots immediately reached for the controls.

Airline pilots rarely perform hands-on, nonautomated flying, especially while cruising in thin air. The autopilot does a much smoother job. This means modern pilots have little practice flying the aircraft at high altitude even though that is where it spends most of its time. The final report on Air France Flight 447 confirmed this reality: "Current training practices do not fill the gap left by the non-existence of manual flying at high altitude."[17] Jolted into action by the sudden disconnection of the autopilot two hours and ten minutes into the flight, the pilots were ill-prepared to manually control the aircraft, especially since they could not see the Earth's horizon due to darkness and clouds. Unaware that the source of the mayhem could be explained by the simple icing up of the airspeed sensors, the pilots sought to maintain control and analyze the problem. They were conditioned to refer to the Electronic Centralized Aircraft Monitoring (ECAM) display to assess the situation, but the ECAM's display gave no warning about ice-clogged pitot tubes and faulty airspeed data. This masked the real source of the problem. The accident investigators concluded that the pilots were "unable to identify any logical link" between what was wrong with the airplane and what the airplane's computer told them on the ECAM display.[18]

In addition to loss of the autopilot, the ECAM display also indicated the automatic flight control system was now functioning in "alternate law" with "protections lost."[19] This meant the aircraft's computers could no longer enforce the preprogrammed automatic limits that protect it from pilot mishandling. The cybernetic system had broken down.

The Airbus A330 is a highly computerized multisystem cybernetic machine that imposes automatic limits on flight control movements to protect itself from pilot mistakes. Its fly-by-wire electronic control system prevents pilots from mishandling the controls and flying it into a stall or exceeding speed, G-force, or roll limits. This ability to automatically override manual control inputs to ensure "complete protection of the flight envelope" is called "normal law."[20] Under normal law, the pilots couldn't stall the semi-robotic airplane even if they tried.[21] The computer would negate even the grossest mishandling by pilots, at least while it operated normally.

In abnormal situations, such as when its airspeed sensors fail, the flight control software reverts from normal law to "alternate law" and imposes fewer limits to human input.[22] And so it was with Air France 447 over the Atlantic. When the electrically heated pitot tubes couldn't melt the ice accumulation fast enough, the A330 lost its ability to sense airspeed and defaulted to alternate law to permit a greater degree of pilot action. In this phase, the automatic protection against pilots exceeding the stall angle no longer existed. The pilots mistakenly assumed the aircraft was going too fast, pulled back on the controls, and quickly flew the Airbus into a high-altitude aerodynamic stall well below the minimum airspeed needed to remain airborne. This was a maneuver they had never seen or done during training and would not have been permitted to do in "normal law." In the degraded state of alternate law, the A330 sensed it was approaching a dangerous stall condition yet could no longer block pilot input. It could only broadcast "stall, stall" over the cockpit loudspeakers along with a distinct cricket-like chirp to try and get the pilots' attention.[23] The pilots ignored these warnings as they unknowingly stalled the aircraft. As it subsequently entered a seven-mile plummet toward the ocean, the pilots sank deeper into a state of disorientation and confusion.

Consider the final minutes of confused dialogue on Flight 447's cockpit voice recorder. The captain, out of the cockpit on his rest break, was summoned back shortly after the crisis began. In his absence, the copilots had climbed over 2,000 feet before stalling the aircraft. It was passing through the initial cruise altitude of 35,000 feet in a nose-high, wings-level, uncontrolled descent when he returned.

2:11:24.6=[copilot in the left seat]: "Do you understand what's happening or not?"
2:11:32.6=[copilot in the right seat]: "I don't have control of the airplane any more now . . . I don't have control of the airplane at all."
2:11:42.5=[Captain, as he opens the cockpit door]: "Er what are you (doing)?"
2:11:43.0=[copilot in the left seat]: "What's happening? I don't know, I don't know what's happening."
2:11:45.5=[copilot in the right seat]: "We're losing control of the airplane there."
2:11:46.7=[copilot in the left seat]: "We lost all control of the airplane, we don't understand anything, we've tried everything."
2:11:58.2=[copilot in the right seat]: "I have a problem. It's that I don't have vertical speed indication."
2:12:01.0=[Captain]: "Alright."

2:12:01.0=[copilot in the right seat]: "I have no more displays."
2:12:02.5=[copilot in the left seat]: "We have no more valid displays."
2:12:04.3=[copilot in the right seat]: "I have the impression that we have some crazy speed, no, what do you think?"
2:12:06.6=[copilot in the left seat]: "No."
2:12:14.4=[copilot in the left seat]: "What do you think about it, what do you think, what do we need to do?"
2:12:15.5=[Captain]: "There, I don't know, there, it's going down."

The next minute includes brief remarks among the three pilots regarding some of the basic flight parameters of the aircraft such as speed, altitude, climb-versus-descent, throttle setting, and roll angle. The sixty-one seconds before impact indicate increasing confusion and exasperation:

2:13:25.3=[copilot in the right seat]: "What is . . . how come we're continuing to go right down now?"
2:13:28.2=[copilot in the left seat]: "Try to find what you can do with your controls up there."
2:13:31.5=[Captain]: "It (won't do) anything."
2:13:36.5=[copilot in the right seat]: "Nine thousand feet."
2:13:38.6=[Captain]: "Careful with the rudder bar there."
2:13:39.7=[copilot in the left seat]: "Climb climb climb climb."
2:13:40.6=[copilot in the right seat]: "But I've been at maxi nose-up for a while."
2:13:42.7=[Captain]: "No no no don't climb."
2:13:43.5=[copilot in the right seat]: "So go down."
2:13:45.0=[copilot in the left seat]: "So give me the controls, the controls to me, controls to me."
2:13:46.0=[copilot in the right seat]: "Go ahead, you have the controls, we are still in TOGA [take off/go around power setting], eh."
2:14:05.3=[Captain]: "Watch out you're pitching up there."
2:14:06.5=[copilot in the left seat]: "I'm pitching up . . . I'm pitching up."
2:14:06.5=[Captain]: "You're pitching up."
2:14:07.3=[copilot in the right seat]: "Well, we need to we are at four thousand feet."
2:14:10.8=[Captain]: "You're pitching up."
2:14:18.0=[Captain]: "Go on, pull."

2:14:19.2=[copilot in the right seat]: "Let's go, pull up, pull up, pull up."
2:14:23.7=[copilot in the right seat]: "[Expletive]! We're going to crash. . . . This can't be true."
2:14:25.8=[copilot in the right seat]: "But what's happening?"
2:14:26.8=[Captain]: "(Ten) degrees pitch attitude."
End of Recording

In the final forty-five seconds the verbal exchange was accompanied by "synthetic voice" commands over the cockpit loudspeakers. These urgent proclamations included eight "Stall" warnings and seven "Pull Up" warnings from the Airbus's robotic voice. In the final ten seconds of its plummet to the ocean, the airplane essentially shouted to its human crew: "Pull up . . . pull up . . . pull up . . . pull up . . . pull up . . . . pull up . . . pull up . . ." Note that seven seconds before impact one copilot repeatedly exclaimed, "pull up." He was in verbal concert with the machine even though he failed to perceive that the aircraft had stalled. All the way to impact, the pilots were pulling back on the controls with the engines roaring at high power in a vain effort to gain altitude. To break the stall, however, would have required an aggressive but counterintuitive push *forward* toward the ocean in order to obtain the crucial airspeed needed to regain aerodynamic lift and fly upward to safety. This never occurred. Man and machine sensed immediate catastrophe, but their integration had completely broken down.

Accident investigators conjectured that the pilot at the controls may not have absorbed the fact the Airbus was in alternate law and may thus "have embraced the common belief that the aeroplane could not stall, and in this context a stall warning was inconsistent."[24] Additionally, the crew may still have relied on the Flight Director's crossbars per normal habit. In the final moments of flight, however, the Flight Director inconsistently displayed the crossbars, and the pilots may have channeled too much attention to the horizontal crossbar.[25] There is a significant possibility the horizontal crossbar indicated a position above the artificial horizon, thus affirming the misassumption the nose should be raised when in fact it should have been lowered to regain airspeed and break the stall.[26] Despite the potentially confusing visual displays provided by the Flight Director, the repeated aural "Stall" warnings were correct. The pilots remained unconvinced of this truth. To add to the confusion, the Airbus's automatic commands to "pull up, pull up . . ." in the final seconds were triggered by its high sink rate close to sea level. Such an input may have validated the misperceptions of the pilots and aggravated the stall.

The post-crash investigation blamed the pilots: "The crew, progressively becoming de-structured, likely never understood that it was faced with a 'simple' loss of three sources of airspeed information."[27] This destructuring resulted from inadequate communication between the copilots, the captain's late arrival in the cockpit, confusion over cockpit indications that did not comport with pilot perceptions, and failure to understand why the autopilot disengaged itself in the first place. The crew had become destructured not only between themselves but also from the onboard systems and from reality as well. The report concluded that "human operators notice and act according to their mental representation of the situation, not to the 'real' situation."[28] This affirms a phenomenon long familiar to pilots. In some ways the situation of Air France 447 was reminiscent of the classic graveyard spiral where pilots are fooled by their vestibular senses and, failing to properly interpret the aircraft's instruments and realize they are in a death spiral, fly the airplane into the ground. Instead of just their own vestibular senses masking reality, Flight 447's pilots became destructured from reality by mental perceptions and misassumptions resulting from a limited, computer-reliant interpretation of the aircraft's actual performance. The human–machine cybernetic system broke down.

In the absence of a specific message on the Electronic Centralized Aircraft Monitoring display stating the initial cause of the problem, the pilots focused not on the origin of the autopilot failure but on the secondary effects portrayed by the cockpit's computers. In a sense, they tried to fly the airplane based on the modern rules and practices of cockpit information management. They may even have falsely assumed the Airbus continued to operate under the flight envelope protections of "normal law." This masked their ability to realize that their 205-ton fly-by-wire aircraft—a marvel of redundancy and sophisticated automation designed to be stall-proof under normal conditions—had actually departed controlled flight in a classic aerodynamic stall. It had entered an ultimately unrecoverable state of low airspeed, low altitude, and high sink rate.[29] The cockpit recorder—later retrieved from the bottom of the Atlantic—showed that the Airbus impacted the water with its nose pitched up an abnormally high 16.2 degrees, a slow forward velocity of a nonflying 107 knots, and a lethal sink rate of nearly two miles per *minute*. Flight 447 was doomed by the human operators' "total loss of cognitive control of the situation" and their failure to bridge the gap between modern cockpit automation and basic principles of flight.[30]

The final BEA report identified numerous faults and recommendations regarding human and machine. Flight 447's copilots received no training in how to prevent or resolve high-altitude stalls, and they were ill-prepared to manually fly the airplane at high altitude even under normal conditions.[31] It recommended the European Aviation Safety Agency mandate regular practice in stall prevention and recovery, even at high altitude.[32] Given the "complexity of modern airplanes," moreover, pilots should receive more training in the potential causes of autopilot disconnection and how to interpret the sometimes-elusive status of automatic control systems.[33] This includes a better understanding of computerized control levels such as normal law and alternate law and their respective flight envelope protections.[34]

The report also faulted the confusing display logic of the Flight Director system since the positioning of the crossbars on the artificial horizon suggested inappropriate control inputs for the pilots.[35] The Air France investigation illustrates that its pilots—trained more as managers of aircraft systems than masters of stick, rudder, and throttles—lost control of the situation when unexpected events confounded their ability to maintain cognitive awareness and engineer a solution. To the end, three seconds before impact, the junior pilot claimed, "This can't be true . . . but what's happening?"

These findings symptomize what some modern aviation experts see as a larger problem. From his headquarters in Alexandria, Virginia, the president of the Flight Safety Foundation recently claimed, "We are seeing a situation where we have pilots that can't understand what the airplane is doing unless a computer interprets it for them." He noted this problem poses a training challenge for the entire industry, not just for Air France or Airbus.[36]

Similarly, Dr. Kathy Abbott, the chief scientific and technical advisor for flight deck human factors for the Federal Aviation Administration (FAA), argues that, while modern cockpit automation overcomes various forms of pilot error, it also introduces new potential errors that may impact how aviators "safely manage the flight path of the airplane."[37] Relying heavily on computer-based flight, pilots experience a new type of problem sometimes called automation surprise. This happens when they are confused or ignorant about what mode the automatic flight control system is in or what it is telling the aircraft to do.[38] The aircraft may do something unexpected or fail to do something the pilot thinks it should; either situation illustrates an underlying problem in the human–machine interface with potentially disastrous results. A 2013 FAA report captured many of these concerns, including the observation that pilots often rely too much on automatic systems and hesitate to intervene in their operation. Pilot skills are also lacking in terms of taking over manually in various phases of flight, as demonstrated in the Flight 447 disaster.[39]

Captain Chesley "Sully" Sullenberger, a senior Airbus captain and famed hero of US Airways Flight 1549's emergency engine-out landing on the Hudson River in 2009, summarizes the basic problems of modern, highly automated commercial aviation. In his view, modern automation technology will not eliminate errors but will "change the nature of the errors that are made."[40] Captain Sullenberger views Air France Flight 447 as an emblem of pilots and technology failing to work together.[41] In highly automated aircraft like the Airbus, for example, Sullenberger notes that pilots are still in charge: they control with their minds—but only occasionally their hands—what the aircraft does. They must possess a "real-time mental model" of reality as they translate their will through various forms of computerized technology. You can turn the aircraft by spinning a dial, inputting a new navigation waypoint, or manipulating the actual controls (even though your control inputs are normally limited by the fly-by-wire system). Yet serious problems emerge when pilots misinterpret what the automation technology may or may not be telling them about reality. Such confusion is aggravated by the complacency and stultifying effect associated with relying on redundant automated systems. Pilots tend to feel "out of the loop" and lose their situational awareness as they manage systems that rarely fail to control the aircraft with more efficiency and precision than humanly possible.[42]

In his 2009 analysis of Captain Sullenberger's successful landing of an Airbus on the Hudson River, author William Langewiesche credited commercial aviation's adoption of fly-by-wire technology to a French engineer and test pilot, Bernard Zeigler. As Airbus's chief developmental engineer in the 1970s, Zeigler pioneered the adoption of fly-by-wire flight controls in commercial aviation by redesigning aircraft for automated control. He recognized that machine-based control optimized aircraft performance and safety by offering precise, stable flight instead of the relatively abrupt, inefficient, or unsafe inputs of human operators. This is something bomber pilots knew since the 1930s when they turned over control to the autopilot during the final leg of the bomb run in order to improve bombing accuracy. Today, Zeigler dismisses criticism from those wary of automatic control: "If you want to fly as [traditional pilots], then go fly gliders, become test pilots, for all I care go to the moon. But flying for the airlines is not supposed to be an adventure. From takeoff to landing, the autopilots handle the controls. This is routine. In a Boeing as much as an Airbus. And they make better work of it than any pilot can. You're not supposed to be the blue-eyed hero here. Your job is to make decisions, to stay awake, and to know which buttons to push and when. Your job is to manage systems."[43]

This may be true under normal conditions plus some abnormal ones. As he guided his engine-out A320 to a safe landing on the water, Sullenberger, by popular acclaim the blue-eyed hero of the "miracle on the Hudson," was nevertheless greatly aided by the stabilizing inputs of fly-by-wire control and flight envelope protection.[44] The National Transportation Safety Board determined that a key aspect of Sullenberger's heroics was his early decision to start the auxiliary power unit. This retained electrical power that kept the aircraft's computers running in "normal law" with full flight envelope protections against overbanking or stalling even though the engines had flamed out.[45] He was a systems manager extraordinaire with remarkable situational awareness. Such an appropriate management of systems preserved precise, stable, computer-enabled control and permitted an effective human–machine interface that prevented disaster. This interface broke down in the ill-fated flight of Air France 447.

In his analysis of modern aviation, Langewiesche argues that incorporating automation and fly-by-wire technology into the airline industry redefined flight by transforming the relationship between pilot and aircraft.[46] He portrays the digitization of aircraft control as "the imaginative rethinking of flight" and credits Ziegler's work at Airbus for overcoming the weak link of pilot control.[47] In reality, flight had been reimagined and redefined *decades earlier* as pilots, engineers, and physicians discovered the limits of human capabilities and developed new procedures and technologies of flight to overcome them. This early transformation of the pilot-aircraft relationship found fuller traction in commercial aviation decades later as digital control became cheaper, more sophisticated, and more reliable. Its modern manifestation in the automated cockpit is ubiquitous. Similarly, the early redefinition of flight as a cybernetic, machine-enhanced phenomenon no longer governed solely by a pilot's innate abilities or inherent limits also permitted a transformation of military aviation. Examples of this transformation are found in the latest fighter aircraft that rely heavily on superior technological systems to see and shoot down other aircraft as well as avoid the most dangerous enemy: the ground.

Military Aviation: Flying and Fighting, Upgraded

In 1945 the Army Air Forces' only five-star general, Hap Arnold, gave this guidance to the head of his Scientific Advisory Group: "What I am interested in is what will be the shape of air power, in five years, or ten, or sixty-five."[48] General Arnold sought to leverage the scientific developments and technological trends of the previous decades to develop an independent, high-tech air force "to fight mechanical rather than manpower wars," even if that meant bombers without pilots.[49] Nearly sixty-five years later the US Air Force's top general tolled the demise of the romantic pilot and reaffirmed the Air Force's preference for a lasting relationship with his mechanical replacement:

> Common to our heritage is the relationship between the aviator and the machine, alone together in the vastness of sky or space. The relationship is etched into our very psyche. It is so powerful an idea that it has attracted the best and the brightest that the world has to offer to our Nation's service. It is these people who made us the service of technological innovation; but today, the evolution of the machine is beginning to outpace the capability of the people we put in them. We now must reconsider the relationship of man and woman, machine, and air. We must question, and ultimately answer, manned or unmanned in combat and support aircraft. We must continue to evolve and embrace the culture of technological innovation which has been our hallmark. We have always, and will continue to use this technological innovation to provide for the security of our nation.[50]

General Norton Schwartz, appointed in 2008 as the nineteenth chief of staff of the Air Force, chose these comments for a conclave of influential senior and retired airmen at the Air Force Association in 2009. It was public acknowledgment of a pilot–aircraft relationship that had been in trouble since practically the first time a pilot in a perfectly good airplane flew into fog (and perhaps crashed), fainted under high G-forces (and perhaps crashed), succumbed to hypoxia (and perhaps crashed), missed an important step in a checklist (and perhaps crashed), or couldn't find his way home (and perhaps crashed).

General Schwartz's 2009 assertion to fellow airpower enthusiasts that the USAF must "reconsider the relationship" of human and machine was old news. Within a decade of the First World War, physicians, engineers, and pilots had already demonstrated the machine's increasing ability to outperform its occupant. His speech could have been written decades earlier by Hap Arnold himself, or perhaps by one of Arnold's contemporaries such as General Curtis

LeMay. In a 1946 memo to General Carl Spaatz, LeMay acknowledged, "in acceleration, temperature, endurance, multiplicity of functions, courage, and many other pilot requirements, we are reaching human limits."[51] Similarly, Lieutenant General Ira Eaker claimed in 1947: "We believe that the Air Force stands at the threshold of a new era. Whereas in the past it has been largely a corps of flying men, in the future, certainly, ten to fifteen years from now, it will be more nearly a corps of technicians and scientists."[52] The same sentiments reverberate across six decades: the traditional reliance on pilot capabilities proved incommensurate with technological innovation and the pragmatic interests of airpower enthusiasts. Pilots may still run the Air Force (General Schwartz and his eighteen predecessors wear pilot's wings), yet the institution's power and adaptability rely also on scientists and engineers unenchanted by the sentimental illusion of the dashing pilot coupled with his machine, in full mastery of its performance. Consider, for example, the pervasive nature of machine augmentation and control—as opposed to pilot-based control—in the newest fighter aircraft.

A 2015 study by the Center for Strategic and Budgetary Assessments (CSBA) illustrates how advances in cockpit technology have altered the most romanticized and legendary aspect of military aviation: air-to-air combat.[53] It concludes that the days of aerial dogfighting have largely receded into history. Unlike their swashbuckling white-scarfed predecessors, modern pilots rarely fight at close range. Their aircraft are still designed to roar through high-G turns and outspeed and outmaneuver their opponent in a tangled duel, yet their onboard sensors linked with sophisticated missiles are designed to destroy enemies long before they are visible to the naked eye. Air-to-air kills now rely more on a machine-based "superior situational awareness" and less on a pilot's sharp vision, fast reflexes, and good aim. The advantage goes not to the pilot with better vision and dogfighting skill, but to the aircraft with longer-range sensors and weapons.[54] This situational awareness comes from systems that can distinguish friend from foe at enormous distances, ideally without the enemy's knowledge. The study's conclusion that "advances in electronic sensors, communications technology, and guided weapons may have fundamentally transformed the nature of air combat" is consistent with a broader redefinition of the pilot's role.[55]

Some may argue that the high-speed, high-G capabilities of modern fighters mean that traditional pilot skills still matter as much as they used to. After all, today's menagerie of sleek, agile fighters such as the MiG-31 Foxhound, F-15

Eagle, F-16 Falcon, F-18 Super Hornet, Eurofighter Typhoon, F-22 Raptor, and F-35 Lightning appear to be built around their pilot, instantly responsive to the lightest touch. He or she straps them on and races off. Yet the way pilots achieve situational awareness in combat scenarios—the way they "fly, fight, and win"—is now governed less by human physical prowess or aggressive maneuvering and more by reliance on long-range sensors, sophisticated software, radar- and infrared-guided missiles, and the competent management of data.[56] This is evidenced by one of the newest fighter aircraft, the F-22.

Pilots of the 90th Fighter Squadron stationed near Anchorage, Alaska, received their new F-22s in 2007. This was a suitable year to transition to the newest fighter aircraft since the 90th was celebrating its ninety-year anniversary. The squadron, one of the oldest in the Air Force, had come a long way from flying Sopwith Camels over the trenches on the Western Front. One wonders what the 90th's original pilots would think of the gleaming machines and complex methods of their modern descendants. Perhaps the one thing they would recognize is the unit emblem, unchanged since the Great War: a pair of dice displaying a natural seven. Wars and warplanes may change, but to fly in the 90th Fighter Squadron has always meant you are in "Pair o' Dice."[57] The squadron's lucky emblem suggests an enduring truth: aviators tend to leave as little to chance as possible and prefer to load the dice in their favor. Better to roll sevens than snake eyes. The F-22's "first look, first shot, first kill" capabilities were designed to win a high-stakes game on the first roll of the dice, long before traditional dogfighting skills need come into play.

The "Pair o' Dice" commander, Lieutenant Colonel Joseph Kunkel, recently described the merits of his new aircraft. "What makes it magic," he explained, is its "stealthiness" and its "integrated avionics, which decreases pilot workload immensely." He also admired "the super maneuverability of the airplane so if you get yourself in a jam and have to dogfight, you can be successful."[58] But in modern aerial warfare the ideal fight does not occur in a close-range dogfight. Instead, Kunkel and his peers seek to start and finish the fight BVR, or beyond visual range. Designed to be difficult to detect by enemy radars, the F-22 can use its own sensors, in some cases assisted by large Airborne Warning and Control Systems (AWACS) support aircraft and their powerful radars, to identify and target one or more enemy aircraft scores of miles away. Its long-range missiles can then home in at supersonic speed and do the work that the 90th's early aviators used to do with their machine guns at close range and worse odds. Pilots like Lieutenant Colonel Kunkel succeed by visualizing the combat space not with paltry 20/20 vision but with the streaming data that fill cockpit displays. The appropriate management of such data lets them strike undetected from ranges unimaginable by the 90th's original pilots and long before they get "in a jam and have to dogfight." According to Kunkel, "everyone wants this jet on their side . . . because of the situational awareness it gives the pilots."[59]

As the CSBA report and Lieutenant Colonel Kunkel illustrate, to control the aerial domain it must be scanned and interpreted by sophisticated avionics rather than by human senses. This new form of situational awareness is one in which a pilot's perceptions and actions are governed to great extent by electronic means. Aerial dueling thus involves the agile management of information far more than the agile maneuvering of aircraft, especially since modern air-to-air engagements typically happen BVR without the sharp-eyed combatants seeing each other.[60] Pilots have learned to fly and fight in a digitized arena portrayed on LCD screens where the aviator with the best situational awareness and the longest-ranged weapons wins. This phenomenon prevails in the Department of Defense's most expensive weapon system of all, the F-35.[61]

Lockheed Martin, the lead manufacturer for the $391 billion F-35 program, claims that modern fighters like the F-22 and F-35 can destroy enemy aircraft from a "standoff" distance.[62] This means they can shoot down an enemy aircraft before it can pose a direct threat to them and, ideally, even before it knows it is under attack. This ability "is redefining previous generation air-to-air tactics" by favoring BVR kills with medium- and long-range missiles. The F-35, however, goes much further in terms of how it redefines the pilot's actions and role in a highly complex system. Indeed, the way aviation engineers have integrated the pilot and the F-35 echoes what Norbert Weiner, the originator of cybernetics, anticipated in 1947: sufficiently advanced sensing systems combined with adequate computing power could yield a cybernetic system of prodigious complexity and performance.[63] For an example of this complexity and performance, consider not just the new plane but the new halo of technology enclosing the pilot's head, the helmet.

F-35 pilots accessorize their wardrobe with a $400,000 helmet.[64] Its primary function is perception, not protection. Typically, pilots look down at the instruments on their control panel or view critical flight information on a transparent head-up display (HUD) mounted on top of the panel. Such methods are out of fashion in the F-35. Sporting the new helmet, an F-35 pilot uses a helmet-mounted display (HMD) on which information about basic flight parameters such as airspeed, attitude, altitude, and heading are projected from within the helmet and onto the visor directly in front of the pilot's eyes.[65] In that sense it is akin to wearing a virtual reality headset, but the pilot can still see the outside world through the transparent visor. Taken alone, this seems just an elaborate but incremental improvement over the HUD. The helmet, however, is designed to do much more.

The new helmet is a sensory and display apparatus that computes information about the environment far beyond a pilot's organic capability to perceive. One of the pilots in charge of an Air Force F-35 training center likens the helmet to "wearing a laptop on your head."[66] In addition to flight instrument data, the inside of the visor depicts images of the airspace and terrain from a 360-degree perspective. As the pilot looks around, she can "see" through the bottom and the sides of the aircraft as if she had X-ray vision, because the helmet is synchronized with external video cameras that offer a commanding view from all angles. Sensors gauge where the pilot (or helmet) turns to look, and computers select the right camera feeds that enable the pilot to scan the ground and surrounding airspace as if she were flying in Wonder Woman's invisible plane. External infrared sensors also expand the spectrum of perception by permitting night vision. The helmet and visor thus combine to multiply the pilot's situational awareness via an elaborately complex cybernetic process. As one F-35 test pilot noted, "You can look through the jet's eyeballs to see the world as the jet sees the world."[67]

The jet sees the world in a whole new way. In the high-tech jargon of the F-35's manufacturer, Northrup Grumman, the aircraft enjoys "omnidirectional autonomous detection, typing, and geolocation" of things the pilot may otherwise never see.[68] It is an information omnivore. From great distances the jet's eyeballs can detect and track surface-to-air missiles, enemy aircraft and their air-to-air missiles, enemy equipment on the ground such as rocket launchers, tanks, and artillery, as well as friendly or neutral aircraft and ground assets along with detailed images of the terrain during day or night. Computer programs discern what may be particularly interesting or alarming and inform the pilot. Although a pilot such as Lieutenant Colonel Kunkel may boast of the F-22's dogfighting maneuverability, Northrup Grumman suggests such old-school tactics are both risky and irrelevant when pilots are cloaked in a sphere of information by a machine that keeps them out of harm's way: "Instead of engaging in a classic dogfight, the F-35 has the option to simply exit the fight and let its missiles do the [high-G] turning."[69]

But what does this mean for the pilot? Although the machinery surrounding the pilot has transformed, the pilot's innate capabilities and limitations are the same as when her goggled predecessors flew wooden biplanes. Classic flying skills and sharp eyes may remain somewhat useful, but the brain inside the helmet needs to perceive, think, and act in a new way. It needs to serve a new regime of data-intensive warfare. Real-time information extracted from all angles around the aircraft and transmitted to the helmet requires the pilot to manage an environment flooded with digitally processed information. She must also determine how to optimize her role in various mission scenarios to fully exploit the aircraft's advanced capabilities. This demands a new approach to aerial combat. It thus requires a new approach to how one trains to fly and fight.

According to Lieutenant Colonel Michael Dehner, a Marine Corps test pilot for the F-35, "you just have to essentially build up those pilots a different way . . . [and] go down a different route that's more or less teaching them to be an information manager." Dehner likens the newest fighter to an "information sponge" that creates an "information hub in the sky."[70] Pilots must think less about flying the aircraft and more about managing information and converting it not just into action but also transmitting it to other aircraft as well as ground forces. Accordingly, the training program for F-35 pilots focuses on developing their skills as information managers; the actual flying of the aircraft takes much less time to learn. Indeed, learning to fly the aircraft is relatively simple, but learning to employ it effectively in a complex combat situation takes far more training. For a new F-35 pilot, therefore, eighteen of twenty-five training flights—more than 70 percent—are flown in a simulator where she can practice using the new suite of advanced sensors and avionics in scenarios difficult to rehearse in the air. According to the F-35 Training System director, to exploit the modern technology of combat aviation means "we are redefining how pilots train."[71]

We have seen how optimizing the operation of civilian and military aircraft resulted in a redefinition of the pilot's role and duties. This is especially clear in terms of the pilot's transition from physical controller to information manager. Similar to the problems experienced by civilian pilots, immersing military pilots in a highly automated cybernetic system exposes problems regarding their organic and cognitive limitations and ability to functionally integrate with the machine.

With their new helmets F-35 pilots are more aware of the world, or at least a virtual display of it. But the system still has bugs. When using its night-vision capability, pilots complain of a distracting "green glow" that impedes their view. Sometimes the digitally rendered image of the external environment is choppy and lacks a seamless portrayal and thus invites confusion. If the imagery bumps around during turbulence, it can become unreadable; and if the flight information lags behind reality, it can induce disorientation and motion sickness.[72] Such problems require faster software that can instantly knit together visual data for real-time synchronization with the pilot's own senses, and subsequent versions of the helmet and software have fine-tuned this form of human–machine integration.[73] Despite modern technological interventions, classic forms of spatial disorientation remain a problem for pilots.

Even modern military pilots familiar with advanced avionics and navigation systems are susceptible to the insidious forms of disorientation that claimed lives since the early years of flight. In 2010, for example, an F-22 pilot was killed when his aircraft struck the ground in a steep dive angle at supersonic speed. Investigators concluded that the aircraft was fully controllable and blamed the pilot's "failure to recognize and initiate a timely dive recovery due to channelized attention, breakdown of visual scan, and unrecognized spatial disorientation."[74] The pilot's "loss of cognitive awareness" was aggravated by an air leak problem and oxygen system malfunction that apparently distracted and, perhaps, impeded the pilot who attempted to pull out of the dive too late—just three seconds before impact.[75] A sophisticated flight control technology called Auto-GCAS, or the Automatic Ground Collision Avoidance System, had recently been tested successfully and would have likely prevented this outcome; tragically, the $147,672,000 F-22 was not yet equipped with this comparatively inexpensive enhancement.[76] Since then, the Air Force decided to install Auto-GCAS in its modern fighters such as F-16s, F-22s, and F-35s.[77] This policy recognizes pilot shortcomings while also challenging their span of control.

Under cooperative development by NASA, the Air Force, and defense industry since the mid-1980s, Auto-GCAS technology now serves as a significant safety enhancement that circumvents human limitations.[78] It is designed to prevent what safety engineers call "controlled flight into the terrain," or CFIT. One of the primary causes of deaths in military and civilian aviation, CFIT happens when an airworthy vessel is flown into the ground due to a pilot's loss of situational awareness.[79] In the USAF's fleet of F-16s, 75 percent of pilot fatalities result from CFIT.[80] A common scenario is a pilot encountering limited visibility conditions during tactical maneuvers, such as pulling up after a strafing run, and then becoming spatially disoriented and unwittingly steering the aircraft back toward the ground. Or a pilot may temporarily lose consciousness during a high-G maneuver and fail to regain his senses before impacting the ground. The Auto-GCAS can prevent such disasters by taking abrupt control of the aircraft and maneuvering it to safe, straight-and-level flight well above the terrain.

The Auto-GCAS works by comparing proximity to the ground with the aircraft's speed, trajectory, rate of descent or climb, and other flight parameters. If the Auto-GCAS calculates an imminent impact, it commands the digital flight control computer to take over, independent of pilot action (or inaction), and initiate an aggressive 5-G climb away from the terrain.[81] The Air Force began installing Auto-GCAS in its fleet of F-16s in September 2014 and predicted this will prevent the loss of at least fourteen aircraft over the service life of the F-16.[82] The effort quickly paid off: in November 2014 during an F-16 combat sortie against Islamic State (ISIS) forces in Syria, the Auto-GCAS intervened to prevent a crash and permitted the pilot to safely return his aircraft to its base in Jordan.[83] Since he avoided the alternatives of being killed on impact or ejecting and falling into the hands of ISIS forces, one surmises he is a fan of Auto-GCAS.

At the point of no return, only 1.5 seconds until impact, the Auto-GCAS seizes control from the pilot.[84] The computer recognizes only what is about to happen, not why it is happening. It cannot calculate if the situation is due to pilot choice, error, or incapacity. It doesn't care. But it is important that the Auto-GCAS not intervene too soon or at inappropriate times and limit a pilot's freedom of action. Sometimes a pilot may need to skim rough terrain at high speed in a combat environment without the risk of the Auto-GCAS triggering an unwarranted climb to higher altitude. To ensure it would still permit aggressive maneuvers, test pilots forced the system to deal with a variety of situations that may occur in full-out combat to see if it limited pilot action. They determined that pilots could still fly extremely low and fast by setting the Auto-GCAS to initiate a fly-up at just fifty feet above the ground. This means the system has enough accuracy and processing power to not be triggered by aggressive combat maneuvers. For an additional margin of control, pilots can set the system into a less sensitive "low protection" mode during low-level strafing maneuvers or flight in mountainous terrain to minimize automatic fly-ups.[85] Such inputs retain a measure of pilot agency even though the Auto-GCAS remains on guard against human misjudgment or incapacity.

Another system augments the Auto-GCAS and gives military pilots an additional option to relinquish control to the flight computer. The Pilot-Activated Recovery System (PARS) enables pilots to command an automatic fly-up maneuver even before one is triggered by the Auto-GCAS. By manually toggling a switch, the pilot can observe as the plane maneuvers itself into safe parameters. This is useful in scenarios such as when the pilot recognizes he is spatially disoriented or somehow unable to ensure safe control.[86] It also adds a second layer of safeguards by not waiting until the literal last second before the Auto-GCAS kicks in. Notably, the pilot still retains the ability to turn the Auto-GCAS off in case it malfunctions. Combined with the PARS, this permits pilots to perform tasks that may activate the system on their terms rather than wait for the system's decision to autonomously intervene. Pilots can thus remain in control, but only if they act before another system component—an automatic one with superior situational awareness and faster reflexes—takes over. But will pilots trust this new system?

University, NASA, and Air Force researchers demonstrated the reliability of Auto-GCAS even under extreme circumstances but still wondered if pilots would deem it trustworthy.[87] They concluded that close communication and mutual respect between aviation engineers and test pilots were vital. Test pilots tended to view engineers as trustworthy and credible when engineers understood the needs and concerns of the pilots. For example, the engineers' primary two principles in designing the Auto-GCAS system for the F-16 were "do no harm" to the pilot or aircraft and "do not impede" the pilot with unwarranted nuisance activations of the system.[88] Like most pilots, the engineers were also skeptical of the system until it proved consistently trustworthy under extreme circumstances. Such attitudes bolstered the confidence of pilots. Once convinced of the reliability of both the system *and* its engineers, most test pilots concurred with engineers on the idea of granting a relatively high degree of decision making to the computer for safety purposes.

The researchers measured pilot and engineer attitudes on a scale of 1 to 10 regarding the acceptable degree of computer control in a human–computer relationship.[89] Originally developed for assessing the level of remote control of undersea vehicles, this widely used taxonomy identifies "1" as total human control until the human decides to give the computer some control. At the other end of the scale, "10" refers to total computer control where the computer may or may not decide to tell the human what it is doing.[90] The value of "7" was selected by 65 percent of the pilots and 80 percent of the engineers: the computer takes control but must then inform the pilot it did so. This relatively high value on the spectrum of machine autonomy suggests a general cultural alignment between test pilots and engineers regarding Auto-GCAS control technology as well as positive results from rigorous field testing. The researchers speculated whether operational pilots would express as much trust and confidence in submitting to Auto-GCAS since they possess a more mission-oriented attitude than test pilots; however, the fact that the system saved one of their own from death or capture by ISIS in Syria may encourage the wider pilot community to accept a high degree of computer autonomy.[91] Such acceptance is indicated by one F-16 pilot's remark about Auto-GCAS: "We like the fact that it's always running in the background and it can save us from our mistakes."[92] But to what degree will military pilots trust computer control?

To further examine this issue, the Air Force Research Laboratory's Human Performance Directorate initiated its own field study in 2015 that evaluates pilot opinions about autonomous systems. The project's goal is to "enhance performance and mitigate factors that undermine trust calibration."[93] This suggests that the problem with pilots goes beyond physical shortcomings such as comparatively slow reflexes or the propensity to become disoriented. The problem includes how to get them to trust and submit to superior forms of control.

A clear example of developing a pilot's trust and willingness to submit to machine capabilities is the necessity of instrument-based flight. Consider the simple mantra favored by an experienced Air Force instructor pilot: *Thou shalt trust thine instruments lest the ground rise up and smite thee.*[94] Pilots have embraced this commandment since instrument-based flight was established in the 1930s; they knew that violating it risked disorientation, professional embarrassment, and perhaps death.[95] Submitting to and trusting a new regime of machine-based control thus possess early precedent. It also aligns with the pragmatic, institutional imperatives to expand aviation's potential and promote safety.[96] The institutional demand to reduce accidents was illustrated by the secretary of defense in 2003 when he mandated a 50 percent reduction in preventable fatalities. This led to the development of the Defense Safety Oversight Council and the Fighter Risk Reduction Project, which provided impetus for the widespread fielding of Auto-GCAS and PARS.[97] These automatic systems, plus the computer-intensive aerial combat capabilities integrated in modern aircraft such as the F-35, add additional layers of technology to overcome pilot shortcomings and displace the human component. Perhaps a second commandment to aviators is needed: *Thou shalt trust thine automatic systems lest the ground or thine enemies rise up and smite thee.*

Combat aircraft provide a rich substrate for interaction and change in the human–machine relationship since they push humans to their limits. This is clearly visible since the early decades of aviation where human shortcomings were exposed and mitigated by the work of physicians, engineers, and pilots to advance the overall capability of the system, even if it altered the pilot's role. Computer-age engineers and airpower enthusiasts have used sophisticated processors and sensors to catalyze even more changes in the pilot–aircraft complex. The pilots of modern fighter and attack aircraft have thus witnessed dramatic shifts in their role and span of control. Beyond Visual Range engagements that rely more on long-range radars and missiles and less on physical dogfighting skills now dominate air-to-air combat; increasingly elaborate sensors multiply situational awareness; sophisticated systems anticipate crashes and decisively intervene to prevent them; and training focuses on making pilots proficient overseers and managers of information. Such changes affirm the directive by the Air Force chief of staff in 2009 to reconsider the human–machine relationship and "continue to evolve and embrace the culture of technological innovation which has been our hallmark."[98] Will modern military pilots accept automation's increasing usurpation of their authority and accept innovations that redefine their relationship with the machine? The history of flight suggests they will.

Conclusion

Both modern commercial and military aviation compensate for the fundamental shortcomings of pilots and require them to reimagine their role. Today's passenger and combat aircraft are safer and more efficient when machines sense the external environment and actuate controls because humans are far more prone to inefficiency and error in these traditional tasks of flight. Pilots thus serve mostly as expert system managers and safety observers. In unusual or emergency situations, this management role may require direct, hands-on intervention. Just ask the passengers on the A320 that Captain Chesley Sullenberger glided onto the Hudson River. His actions were crucial, although the fact that his control inputs were still governed by flight control software indicates the pervasive role of machine-based control. Despite such advances, the pilot's prevailing role as a manager of complex control technology creates its own risks. As demonstrated in the Air France 447 disaster, the manual control skills of modern pilots may atrophy, and, crucially, it is easy for pilots not only to become complacent but to misunderstand the nature and extent of computer-based control. As Sullenberger noted in 2009, this pilot–machine interface changes "the nature of errors that are made." Changes in the nature of errors reflect a change in the nature of flight.

These changes illustrate the role of trust in aviation. Pilots have been increasingly required to entrust their own lives and, in the case of military pilots, their combat effectiveness to the demonstrable superiority of machine-based situational awareness and precision control. At the same time, a century-long effort to improve the safety and performance potential of aircraft has taught aviation leaders, engineers, physicians, and pilots that the man or woman at the controls cannot be fully trusted due to the problematic limits of human performance and cognition. Thus, the efforts to recalibrate a pilot's duties as well as his or her willingness to trust sophisticated forms of automatic control have served the pragmatic interests of civilian and military aviation, even if it undermines the traditional agency and prestige of the pilot.

The long trend of shifting trust from human to machine has presented new roles and challenges for pilots as they cling to their traditional image as prime arbiters of control. Given the leaps in processing power and automation software, the obvious question is whether pilots will be around much longer and, if so, in what capacity?

New Horizons of Flight

The cockpit is an arena where humans have grappled with the extremes and complexities of technology. The arena's boundaries and rules frequently change, posing new challenges and opportunities for those inside and outside the cockpit. Considering how the technoscientific visions of aviation leaders related to the pilot's evolving role is one way to assess those changes.

A foundational 1945 document, *Toward New Horizons*, envisioned a new Air Force suffused by "scientific progress" with a trajectory toward greater automation and even pilotless aircraft. The Air Force's publication of *Technology Horizons* in 2010 and *Autonomous Horizons* in 2015 were rooted in this early vision. They forecast an aggressive pursuit and application of technoscientific advances that recharacterize how (and where) pilots interact with their machines. This includes treating the pilot and the cockpit's autonomous technology as partners, each with unique strengths. As machines become more capable, this may not be an equal partnership. Air Force and Navy leaders have also used technological advances to catalyze the rise of drones, suggesting unmanned aviation may become the "new normal." Such changes require pilots to think beyond the established paradigm of flight as they peer over the horizon toward their own future.

"The Next Air Force Is Going to Be Built Around Scientists"

No one knew why the first airplane flew. The Wright Brothers' triumph at Kitty Hawk occurred without a scientific understanding of flight. Knowledge of how a wing generates lift did not exist until 1904, the year after the first flight.[1] Additional research into airflow principles aided the development of the new science of aerodynamics, yet this knowledge arose only after aircraft took to the skies in America and Europe. The science of flight was thus not a prerequisite for the reality of flight, and we have seen how aviation technology generated new knowledge in various fields of science. The belief that "science discovers, technology applies" is thus a myth.[2] Nevertheless, even though the Wright Flyer affirmed that technology is not necessarily the result of applied science and can appear in the absence of science, aviation's development over the next five decades demonstrated how new scientific knowledge of human and machine dramatically enhanced the performance of the technological system. Aviation evolution thus involved a complex and complementary interaction of science and technology. Airpower enthusiasts developed a firm grasp of this notion.

"For twenty years the Air Force was built around pilots, pilots, and more pilots," declared General Henry "Hap" Arnold, commander of the US Army Air Forces in World War Two, "[but] the next Air Force is going to be built around scientists—around mechanically minded fellows."[3] General Arnold's 1944 statement reflected the traditionally central status of pilots and the desire of airpower enthusiasts to expand the technoscientific mastery of the air. His statement is somewhat misleading, however, since it suggests aviation ignored science in prior decades. In fact, the opposite is true. Despite emerging with little scientific basis, military and commercial aviation cultivated substantial scientific knowledge in the early decades of flight. Physicians and engineers—scientists and mechanically minded fellows—applied this knowledge to establish new roles and responsibilities for pilots, requiring them to adapt to machine advances. While pilots still appeared to play a central role, innovative research and engineering permitted the ambitions of airpower enthusiasts to be scientifically informed and not limited by the capabilities, traditions, or even presence of human pilots.

General Arnold, the top air force officer since 1938, provided much of the impetus toward pursuing new research and technological devices that often ignored or superseded the role of pilots. A glimpse into his mindset comes from a March 1941 meeting regarding the development of bombs capable of gliding great distances. An observer noted, "General Arnold, with his customary hammering on the table, stated that action wasn't happening fast enough and insisted that we get a gray-bearded scientist and start working on it."[4] Arnold continued hammering away at the importance of seeking new ways to master the air, resulting in the wartime development of pilotless weapons such as the unmanned Weary Willie B-17s and the "Jet Bomb 2" (JB-2) along with radar- and radio-based control and other automation-related technologies that usurped pilot mastery.

The connection between military goals and the scientific community was symbolized in the relationship between General Arnold and Dr. Theodore von Kármán. A physicist and aeronautical engineer, von Kármán emigrated from

Budapest in 1930 at the behest of Robert Millikan, a physicist and Nobel laureate from the California Institute of Technology. Millikan installed von Kármán as the head of Caltech's Guggenheim Aeronautical Laboratory, the same organization that conducted the first blind flight in 1929.[5] General Arnold met Dr. von Kármán in 1936 and established their mutual interest in modernizing American aeronautics. One of their contemporaries, Lieutenant General James H. Doolittle, described their relationship in these terms: "While General Arnold was not a highly technical man he did understand the importance of science and technology, and while Dr. von Kármán was not strictly a military man he realized the importance to the military of mobilizing science."[6] In 1939, for example, Arnold shared his belief with von Kármán that the Air Corps must devote more resources to experimental work in order to improve its stable of aircraft; von Kármán then agreed to design a twenty-foot wind tunnel, the largest of its kind, for research at Wright Field.[7]

Confident in von Kármán's reputation and ability to get things done, Arnold arranged a meeting with him in September 1944. "Only one thing should concern us," the senior-ranking general declared: "What is the future of air power and aerial warfare?"[8] General Arnold wanted Dr. von Kármán's help in designing the future Air Force and asked him to create a science-based blueprint for the next twenty years. A new organization, the Scientific Advisory Group, emerged from this meeting with von Kármán at the helm. His belief that "future warfare would have a primarily scientific character" fit with General Arnold's vision, and he recruited men from industry, academia, and the military who could advise senior military leaders "on the relations between scientific thought, technical research, and air power."[9] When the Scientific Advisory Group, later the Scientific Advisory Board, met for the first time in late 1944, General Arnold emphasized that he wanted its members to "search into every science to squeeze out basic developments that could make the United States invincible in the air."[10]

The relationship between General Hap Arnold and Dr. Theodore von Kármán and their creation of the Scientific Advisory Board reflected the importance of science to the development of military aviation.[11] Yet scientific experimentation had already become part of the sinew of airpower during the preceding decades of flight, and the scientists and engineers associated with the early years of military and civilian aviation set the stage for new visions of airpower in World War Two and beyond. Von Kármán's 1945 publication, *Toward New Horizons*, captured this sentiment by declaring the next decade should be devoted to "scientific progress" and should, among other goals such as supersonic flight and improved communication, include "pilotless aircraft" and "remote-controlled and automatic fighter and bomber forces."[12]

The vision for future airpower put forth by *Toward New Horizons* in 1945 placed considerable emphasis on developing unmanned aircraft. A key priority in the next conflict would likely be the destruction of enemy atomic weapon production installations, and this would be conducted by sending forth piloted and pilotless aircraft.[13] Von Kármán noted that "the high speed of pilotless airplanes and missiles" would thwart enemy defenses, and advances in gyroscopes and automatic pilots would enable the remote-controlled strike of different targets.[14] Such predictions proved to be premature. Although pilotless aircraft developed a niche role in the second half of the twentieth century, the projection of power via unmanned means was manifested primarily in the form of cruise and ballistic missiles. Von Kármán's assumption in 1945 that automation technology and remote control would completely solve "the problem of pilotless aircraft" exceeded scientific and technological realities.[15] Instead, increasingly accurate ICBMs satisfied much of the institution's need to project power at great distance to help deter superpower conflict.

The notion that airpower enthusiasts, many of them pilots, would seek to expand control of the air through highly automated pilotless technologies indicates that preserving a pilot's traditional role, status, and span of control was not an essential requirement in aviation development. In addition to new means of automatic control, the aviation system had undergone vast increases in scale and complexity. It was now a massive, evolving ecosystem of airfields, navigational aids, radar and communication technologies, training and evaluation procedures, experimental engineering, maintenance depots, government and private research entities, military doctrine, safety regulations, fuel infrastructure, and the growing network of commercial travel.[16] As this system expanded it demanded reliability, predictability, and efficiency from all of its components, including pilots. Pilots accordingly became more like interchangeable cogs in a system governed by logic and standardized performance that could push them to their cognitive and physiological limits.[17] In that sense they had become servants to technology rather than masters of it.

Yet in 1945 pilots were in little danger of being physically evicted from most cockpits. Even in his pragmatic, scientific approach to aviation's future, von Kármán acknowledged an ongoing need for manned aircraft. Automation devices could not yet substitute for a human's ability to react in the dynamic, unpredictable environments that characterized many aspects of aviation. He thus believed unmanned aircraft would not entirely supplant manned ones and envisioned aviation as the ongoing effort to improve performance by overcoming the natural limits of humans.[18] This necessity to overcome human limits, of course, animated the earlier development of instrument flight, pressurized cabins, and automatic flight controls and redefined the pilot's role. This legacy endured as pilots increasingly served as managers of technological systems whose functions far exceed human capabilities. Consider the modern versions of 1945's *Toward New Horizons* recently promulgated by Air Force leaders.

Augmentation and Autonomy

In 2009 the secretary of the Air Force and the chief of staff of the Air Force directed their chief scientist to create a twenty-year forecast of the scientific and technological developments that will shape the airpower institution. They sought to ascertain "leapfrog" and "game-changing" capabilities that would create new challenges and opportunities for command of the air. Noting the Air Force's historic reliance on scientific and technological advances, they directed scientists, engineers, and other experts from within and outside the institution to collaboratively create "the world's most capable Air Force."[19] This directly echoed earlier generations of airpower enthusiasts who sought a broad spectrum of means to expand institutional power.

The resulting 2010 study, *Technology Horizons*, supplied a forward-looking assessment of aviation evolution that depicted a fundamental role for autonomous technology and affirmed that humans represented the weakest component in many aspects of aviation. A key element of the report involved the ability and willingness to trust highly adaptive automated systems. In its recommendation to replace human functions "in whole or in part," the study acknowledged the need to verify and validate the effectiveness of such systems in order for pilots to deem them reliable and trustworthy.[20] Verification, validation, and subsequent changes in pilot attitudes thus constitute key aspects of technologies that permit higher levels of autonomy, safety, and effectiveness.

The willingness of pilots to significantly change how they function in flight has ample precedent, such as the transition to instrument flight during the interwar years. To expand the limits of flight, military and commercial pilots learned to submit their own physiological senses to the superiority of instrument-based control. Professional pilots internalized the mandate to trust instruments instead of themselves, and it became increasingly common to link instrument-based control with the use of automation technology. Enhancing overall system performance thus involves a process of trust calibration to generate acceptance and incorporation of superior forms of control.[21] In addition to automation improvements, *Technology Horizons* also viewed performance enhancement from a different angle: "direct augmentation of humans."[22]

We have seen how augmenting human capabilities underlies much of aviation's history. Pilots were outfitted with devices such as oxygen masks, G-suits, heated clothing, and pressure suits to withstand the extremes associated with the expanding performance envelope of aircraft. Sometimes they improved their mental alertness via amphetamine-laced supplements. Yet such augmentations are insufficient. The Air Force Research Laboratory's Human Effectiveness Directorate currently works to develop a more sophisticated array of devices to sense the physical and cognitive state of pilots during the stresses of flight. These devices provide real-time assessments of human physiological cues, such as eye movement, cardiac function, voice inflections, facial expressions, thermal signatures, temperature, respiration, hormonal and neurochemical levels, and brainwave patterns.[23] The goal: establish the physiological and neurological metrics to assess the limits of human performance and how various augmentations may expand such limits. These efforts reflect the recommendation of the 2010 *Technology Horizons* report to address the growing mismatch between the relative constant of basic human capacity and the steep acceleration of machine capability, a phenomenon Hap Arnold recognized in 1925.[24]

Keeping up with technological capabilities involves more than altering the pilot's role, it may involve altering the pilot herself. To exploit the tide of data flooding into the cockpit, the report recommended the use of implants and drugs to boost "memory, alertness, cognition, and visual/aural activity."[25] Feasibly, this could involve interfacing humans and machines via brainwave activity. Just as pilots are screened for visual acuity, future pilots could be selected on their ability to sync their brainwave patterns with external devices. The report even suggested the possibility of genomic screening and genetic engineering to identify and improve such capabilities.[26] Such developments, however distant in the future, cohere with the general effort to achieve "synthetically augmented intuition" that improves the speed and quality of decision making.[27] Just as interpreting data from cockpit instruments provided the synthetically augmented intuition necessary for Jimmy Doolittle to fly the first all-blind sortie in 1929, modern aviation enthusiasts envision new levels of performance via bioengineered, machine-augmented human operators in those situations where humans may still be required.[28]

In addition to altering the pilot to keep up with the demands of her aircraft, an alternative is to replace her with intelligent machines. In that vein, and in a nod to 1945's *Toward New Horizons* and 2010's *Technology Horizons*, in 2015 the Air Force Office of the Chief Scientist published *Autonomous Horizons: System Autonomy in the Air Force—A Path to the Future*. The report invokes the term *autonomous* instead of *automatic* to reflect advances in software and machine intelligence.[29] Unlike automatic control systems designed to function within specific parameters and in predictable situations, autonomous systems possess a higher order of machine intelligence that permits adaptation and decision making in dynamic, unprogrammed environments.[30] The report thus applies the leaps in computational power in the age of Moore's Law to the long tradition of augmenting pilots and displacing pilot control with machine alternatives.

In its effort to chart the new roles of modern pilots, *Autonomous Horizons* diagnoses a primary malady of automation: brittleness. From early devices such as gyroscopically based autopilots to later systems such as the Airbus automatic flight management system, automation technology is brittle—it works well under carefully programmed and predictable conditions, but it fails in situations of unanticipated variables and stresses.[31] Recall the practice of "V-1 tipping": British pilots could destroy German V-1 cruise missiles by using their wingtips to tip them over and thus tumble the V-1's primitive gyro, causing it to spiral from the sky. The V-1s were brittle, able to operate only in predictable, stable circumstances. To reduce brittleness and enhance automation's flexibility and resilience, the 2015 report prescribes the improvement of machine intelligence to transition from automatic to autonomous technology. Autonomous systems—referred to simply as *autonomy*—can then form a dynamic, evolving partnership with human operators.[32]

The Air Force chief scientist surmises that improvements in computing power and machine intelligence, perhaps via advances in neural networks and learning algorithms, will broaden the spectrum of machine control. Unmanned aircraft, for example, could achieve semiautonomy or even full autonomy in a greater number of operational scenarios when they are governed by more sophisticated software.[33] Such machines could make decisions previously left to human operators. Presumably this could multiply military options just as instrument-based flight vastly expanded the parameters of manned flight. Similar to Theodore von Kármán's acknowledgment in 1945 that pilots would remain in many types of aircraft, the chief scientist's report asserts that current and future systems will still require some form of interaction with airmen either inside or outside of the cockpit. Yet we have seen that airmen also suffer from "brittleness."

While pilots add flexibility to the overall system, their innate physiological and cognitive limits make them rather brittle creatures prone to failure and dysfunction when thrust outside of fairly narrow operational parameters. Their ability to swiftly and accurately process masses of data or maintain cognitive control of complex systems can be quickly overwhelmed.[34] This is one reason the F-35 is remarkably stable and easy to fly. Its pilot does not have to spend much cognitive bandwidth on maintaining basic control parameters, and this lets her pay more attention to the digital information flooding into her helmet.[35] Such information, however, is extensively filtered and adjudicated by software to appear in a format and volume legible to the pilot. Brittle in different ways, pilots and autonomy must therefore be integrated on the basis of their relative strengths and weaknesses.

Autonomous Horizons portrays the integration of pilots and autonomy in personal terms. They are "collaborating partners," a "team."[36] It promotes the (increasingly) intelligent machine to the status of partner. The dynamics of this partnership can vary based on circumstance and machine ability, and the report identifies advancing levels of autonomy. These range from "manual control" (all actions performed by pilot) to "decision aiding" (system provides analysis and recommended options to pilot) to "full autonomy" (system exerts complete control independent of human guidance).[37] As discussed in chapter 6, this spectrum coheres with the long-established taxonomy of human-versus-computer control developed by MIT engineers that includes numerous intermediate steps between the extremes of human-only and computer-only control.[38] The extent of collaboration between human and machine systems, plus the degree of autonomy pilots may be willing to tolerate, depends on an underlying element of most successful partnerships: trust.

Partner relationships evolve through familiarity and trust. The symbiotic pilot–machine relationship is shaped by the pilot's confidence in autonomy and his ability to verify its performance and judgment. If deemed reliable in basic as well as complex scenarios, autonomy's "span of control" can widen.[39] This manifests in the study of trust calibration by the Air Force Office of

Scientific Research. Its researchers seek to enhance the role and acceptance of autonomy by examining pilot trust in automatic ground collision avoidance systems (Auto-GCAS) that can wrest control away from the pilot when necessary, as noted in the previous chapter.[40] Apparently Air Force leaders trust such systems since they ordered their installation in the F-16, F-22, and most recently the F-35.[41] They are also testing the trustworthiness of a new technology, an automatic *air* collision avoidance system, that executes aggressive maneuvers to avoid other aircraft and "automatically locks out the controls [from the pilot] until danger is averted."[42] Consistent with equating pilots and autonomous systems as trusting, collaborative partners, *Autonomous Horizons* notes that this doesn't involve just the human perspective. Trust works both ways.

Just as pilots evaluate autonomous systems, these systems may also verify if the pilot is trustworthy. Autonomy may assess and manage its human partner. Is the human component functioning adequately? Is the pilot making good decisions?[43] As the pilot-autonomy partnership matures, autonomy can potentially improve its own situational awareness by evaluating the competence, dependability, and timely actions of the pilot. The aircraft's sensors can probe the external environment *and* the human operator (temperature, blood oxygenation, eye movement, facial expressions, brainwaves, etc.). Simultaneously, the aircraft can compare the pilot's status and actions to its own interpretation of the environment. By evaluating the pilot's decisions and measuring her physiological and cognitive status, autonomy can aid their "shared situation awareness" to optimize overall performance.[44] The Air Force report predicts that both partners will learn to perceive what the other wants to do and team up to play on each other's strengths.[45]

We have seen the pilot transition from primary operator to systems manager, and *Autonomous Horizons* envisions a new transition for the pilot from systems manager to codependent partner. She will partner with a form of autonomy able to develop its own situational awareness, evaluate various courses of action, and execute those actions in concert with her or, in some scenarios, independent of her action or inaction. Sometimes the pilot will be "in-the-loop" playing an active role in system operations; sometimes she will be "on-the-loop" observing autonomous operations yet able to intervene; and sometimes she will be "out-of-the-loop" and unable to intervene, relying on total autonomous control in emergency, time-critical situations, such as unconsciousness or insufficient situational awareness of impending threats to the aircraft.[46] In this modern paradigm of aviation, pilot and autonomy are codependent partners, both able to assess and to accommodate each other's limits and strengths.[47] It is an advanced form of human–machine complementarity.[48] This partnership, however, does not necessarily rely on a physical coupling and eases the displacement of the pilot from the cockpit.

Unmanned Systems and the New Normal

On 15 April 2015, Secretary of the Navy Ray Mabus offered his vision of the future of naval aviation. He declared that the F-35—commonly called the "Joint Strike Fighter" since it is a joint effort of the Navy, Air Force, and Marines—will "almost certainly be the last manned strike fighter aircraft the Department of the Navy [including the Marine Corps] will ever buy or fly."[49] In subsequent remarks to the National Press Club, Secretary Mabus praised the capabilities of the F-35 and noted that it makes decisions faster than its pilot can. Yet he described unmanned systems, not highly sophisticated man–machine complexes such as the F-35, as "the future of warfare." Requirements for greater endurance and risk minimization, plus the institutional imperative to "keep up . . . because we're not the only ones working on this," may establish unmanned, autonomous systems as the "new normal" in much of military aviation.[50]

In the pursuit of a new normal, Secretary Mabus fostered the development of an experimental aircraft for the Navy: the unmanned carrier-launched airborne surveillance and strike aircraft (UCLASS), or X-47B. A fully autonomous machine, in July 2013 the X-47B demonstrated the ability to land on an aircraft carrier—arguably one of the riskiest maneuvers in aviation—without any human input. The footage is surreal: a machine roughly the size of a manned fighter aircraft, except with an engine air intake where one might expect the cockpit, making a perfect touchdown on a moving aircraft carrier every time it tried. When asked by a skeptical pilot if the X-47B can successfully execute carrier operations at night, a program engineer replied: "Sir, we haven't done any testing at night. Again, the aircraft is a robot. It doesn't have eyes, and can't tell whether the sun is up or not. It's autonomous. There is no pilot. It can't see because it doesn't have to."[51]

In addition to landing on an aircraft carrier (day or night), the UCLASS aircraft also demonstrated the ability to conduct another high-finesse maneuver, aerial refueling, without human input. For its advances in aviation autonomy, the X-47B program won the National Aeronautical Association's 2013 Robert J. Collier Trophy, a prestigious annual award recognizing "the greatest achievement in aeronautics or astronautics in America with respect to improving the performance, efficiency, and safety of air or space vehicles."[52] This achievement was the latest in a long lineage of Collier winners who advanced various elements of machine augmentation and automation, including: Orville Wright (1913, automatic stabilizer), Elmer Sperry (1914, gyroscopic control), Albert Hegenberger (1934, the first blind-landing system), Luis Alvarez (1945, radar-based ground control approach systems), and William Lear (1949, automatic pilot improvements).[53] These earlier achievements did not lead inexorably to the autonomous X-47B, but they represent some of the foundational aspects of machine-based control that paved the way for what is in essence a flying robot.

The early developments in augmentation and automation permeated aviation and reshaped the pilot's role, and there is a similar phenomenon in modern aviation as unmanned technology seeps into modern aviation. The Navy announced in February 2016 that it will transition the X-47B UCLASS into a new system: the carrier-based air refueling system (CBARS).[54] This will limit and perhaps replace the current and inefficient practice of using manned aircraft to refuel other carrier-based aircraft. Using the X-47B as an unmanned tanker aircraft marks a significant shift from its original portrayal as an autonomous reconnaissance and strike platform. However, according to Chief of Naval Operations Admiral John Richardson, doing so provides the swiftest alternative to integrating unmanned aircraft into carrier-based operations "with a legitimate role to play because there's just so much to learn right now."[55] The Navy claims the new CBARS aircraft may even include a modest reconnaissance or strike capability, but the priority is to develop basic expertise and a community of practitioners who can foster additional developments in autonomous aviation.

Encouraged by the success of the X-47B UCLASS, in 2015 Congress added $350 million to the Navy's original request of $134.7 million "for continued development and risk reduction activities" and to hasten the integration of manned and unmanned aircraft.[56] Congress also authorized drone pilots $150 more in monthly flight pay and $10,000 more in annual bonuses than their counterparts operating manned aircraft.[57] This means pilots controlling aircraft from remote consoles may now earn more than their counterparts who actually strap into aircraft. Unsurprisingly, a few months after Congress clearly showed its desire to bolster the development of unmanned aircraft, the chief of naval operations remarked, "I'm eager to get started in unmanned technologies."[58]

The Navy secretary's prediction that unmanned, autonomous systems will represent the "new normal" finds evidence in the Navy's recent adoption of a land-based unmanned aircraft that the Air Force has deployed since 2001.[59] The Global Hawk, renamed "Triton" in the Navy, emerged from the Defense Advanced Research Projects Agency's (DARPA) 1994 sponsorship of a high-altitude "advanced concepts technology demonstrator."[60] With its impressive wingspan and powerful engine, the Global Hawk was designed for long-duration cruise above 60,000 feet with a payload of reconnaissance equipment that streams data via satellite to users on sea or land. It is thus ideally suited to conduct surveillance over broad areas of maritime interest during lengthy flights that far exceed the limits of human endurance. This means the Navy will soon operate land-based and carrier-based unmanned systems and strive to integrate these systems in its established aviation culture.

Part of the Navy's motivation, as suggested by the secretary of the Navy's comments in 2015 about not wanting to be left behind in the arena of unmanned aviation, involves keeping pace not just with potential adversaries but also with the Air Force. In the late 1990s, and particularly after 9/11, the Air Force rapidly expanded its development and use of unmanned aircraft. The term drone is fashionable in common parlance but finds little attraction in the Air Force since it implies a mindless worker. For a similar reason, the Air Force also replaced unmanned aerial systems (UAS) with remotely piloted aircraft (RPA) to emphasize that unmanned systems are still governed by human inputs. For simplicity, this analysis treats the terms drone and RPA as the same.

In addition to developing the Global Hawk in the 1990s, the Air Force fielded the unarmed Predator drone in 1996 as a surveillance platform. One pilot portrayed the slow, propeller-driven Predator as an unmanned version of the Wright Flyer—a new technology brimming with potential.[61] This was an apt comparison. The ungainly airplanes developed in the first decade of the twentieth century initially served as aerial observation posts but were soon armed with means to strike ground targets. Similarly, in the first decade of the twenty-first century, the Predator graduated from a reconnaissance platform to an armed drone with a lethal sting when equipped with Hellfire missiles. A spectrum of new possibilities emerged. The Predator was soon followed by the Reaper, a faster and more sophisticated aircraft capable of carrying heavier payloads greater distances. Within a decade sixty-five Predators or Reapers were airborne at any single moment and mostly on armed patrol in the Middle East and Horn of Africa. They became the weapon of choice for airpower enthusiasts and political leaders determined to prosecute the war against extremist groups, and their elevated status threatened the status quo of conducting strikes from manned aircraft.

The success of drones, or RPAs, generated new expectations despite the challenges they posed to traditional pilot-centric norms within the Air Force. Secretary of Defense Robert Gates directed the Air Force in early 2008 to refocus its views about which types of missions need traditional pilots. Annoyed by the Air Force's delays in satiating the growing appetite for RPAs in Iraq and Afghanistan, he claimed "it's been like pulling teeth" to change entrenched attitudes and accelerate change. Apparently, he had neither forgotten nor forgiven the Air Force's failure to collaborate with the CIA in 1992 in unmanned drone development because this meant "a vehicle without a pilot."[62] Secretary Gates made his comments to students at the Air War College, a cohort of officers likely to populate the most senior ranks of the Air Force. He advised them to rethink "long-standing service assumptions and priorities about which missions require certified pilots and which do not."[63]

On the heels of this criticism the secretary of the Air Force declared in 2008 that RPAs will not be a "one war wonder" and predicted drones would soon constitute 25 to 50 percent of new aircraft.[64] Subsequent years affirmed this forecast. In fiscal year 2012 the number of aircrew trained to fly manned fighters and bombers was eclipsed by those trained for RPAs.[65] By 2014 the Air Force's fleet of Predator and Reaper drones amounted to nearly 400. This was more than the A-10 (359) or F-22 (183), nearly the number of F-15s (468), and approached half of the ubiquitous F-16s (1,017).[66] Accordingly, the number of RPA personnel ballooned from 2,100 in 2005 to an estimated 9,900 in 2017.[67]

Although the pilot–aircraft relationship had already transformed over previous decades to increasingly favor machine-based control, many clung to the belief that manned aircraft would continue to play the lead role in Air Force missions. Yet the grip of this traditional standard was loosened by the rise of unmanned systems. Their demonstrable battlefield success and future potential, combined with changing strategic priorities and pressure from senior civilian leaders to obtain maximum effects with minimum risks, introduced new expectations for military aviation. Moreover, official Air Force reports such as the 2010 *Technology Horizons* and the 2015 *Autonomous Horizons* indicate a burgeoning enthusiasm regarding the future of unmanned—and perhaps autonomous—aviation. Airpower enthusiasts would thus need to consider how drones redefined the role of modern pilots.

"Get Over Your Paradigms"

Unmanned flight imposes a new worldview on aviators. The traditional paradigm of manned flight places the pilot at the center of the aviation universe. Even when not asserting hands-on control and acting instead as an observer, director, and systems manager, the pilot is still the perceived focal point of system function. Pilot identity orbits around this assumption. Modern technology upends this pilot-centric view of flight by displacing the pilot and locating a new center: the machine.

Although a pilot may retain control over an unmanned aircraft, his physical absence demands new ways of thinking about the complexities and possibilities of modern flight. The pilot, now separated from physical risk, manages various subsystems and is more subject to the personal and simultaneous intervention of others such as mission managers, intelligence officers, lawyers, engineers, weapon experts, and superiors. He is no longer the sole point of reference and control. He is immediately interchangeable with another pilot, especially when circumstances such as long-duration missions dictate. In such ways, as one experienced pilot noted, the RPA is a "disruptive technology" that compels a different worldview and a new way of reasoning.[68]

Historian Peter Paret observed that a fundamental challenge of innovation is developing "intellectual mastery" over a new technology.[69] We have seen how pilots sought intellectual mastery over technological changes that expanded the envelope of manned flight. Sometimes this involved new skills such as instrument-based flight; elsewhere it demanded pilots develop a cognitive awareness of machine-based control. Relinquishing more control to onboard systems required pilots to master a new understanding of flight. Unmanned systems such as the Global Hawk, Predator, and Reaper imposed their own challenges in gaining intellectual mastery over a new paradigm of flight that redefined the roles of human operators. Consider, for example, how the Global Hawk requires pilots to think and act in unconventional ways.

The unmanned Global Hawk high-altitude reconnaissance platform still requires several pilots to operate it. Their functions, however, are dramatically different from their peers flying manned aircraft. When a Global Hawk flies a high-altitude reconnaissance mission somewhere over the globe, it is operated via satellite from a control console in the United States by a Mission Control pilot. He uses a keyboard instead of a control stick and is unconcerned with many traditional piloting skills such as takeoffs, instrument-based flight, holding patterns, radar vectors, or landing. Instead, he controls the aircraft by programming navigational waypoints relevant to reconnaissance targets. He does not ponder the question constantly asked by all pilots who fly single-engine aircraft: "What will I do if the engine fails right now?" He wonders instead, "What will the *computer* do if the engine fails right now?" If its engine flames out, the Global Hawk's computer algorithms independently take control and determine the most suitable landing field within gliding distance. All the pilot can do is notify air traffic controllers and ground personnel that a large robotic aircraft will soon line up on a runway of its choice, land, and roll to a stop precisely on centerline.[70]

Despite this high level of autonomy, the Global Hawk is nevertheless subject to some degree of control by two other pilots who constitute the launch and recovery element (LRE). They are based at the Global Hawk's takeoff and landing site. One LRE pilot sits in a control trailer typing in computer-based commands. This pilot cannot see the Global Hawk but can view its flight path through an infrared camera mounted in the aircraft's nose. A second pilot, called Hawk Eye, follows the Global Hawk down the runway and serves as a safety observer. As the Global Hawk taxis on and off the runway, Hawk Eye monitors ground activity and, if necessary, can hop out of the car and shut down the aircraft using its external fuel-cutoff switch.[71] At least three pilots, therefore, are used to operate this semiautonomous machine, yet their roles are far from the ones ingrained into the pilots of manned aircraft. Other unmanned systems, such as the Air Force's Predator and Reaper drones, also challenge traditionally trained military pilots to think and function in substantially different ways.

To understand the fundamental differences between manned and unmanned flight, consider the remarks of two experienced fighter pilots who commanded Predator and Reaper squadrons. One, formerly in charge of the USAF's 11th Reconnaissance Squadron that is responsible for training Predator operators, urged his new pilots to "take out your sledgehammers and start breaking down the walls." To effectively operate Predator drones required pilots to ignore much of their previous training in manned aircraft, so his basic message to pilots was "get over your paradigms."[72] To succeed as aviators they needed to develop a new worldview to gain intellectual mastery over these aircraft.

His peer, the commander of the 42nd Attack Squadron, which operated the more heavily armed Reapers, acknowledged that the squadron's best pilot was actually not a pilot; instead, he was previously qualified as a combat systems operator in the F-15E and never went to pilot training.[73] In this new world, traditional piloting skills were no longer essential and could even be a liability. Pilots assigned to the Predator and Reaper program had to redefine their view of flight.

Learning how to fly a Predator or Reaper drone requires forsaking an ability fundamental to pilot identity: manual stick-and-rudder skills. Although such skills may atrophy in modern aircraft that operate extensively under automatic control, they are still a cherished measure of pilot prowess. Perhaps this is why the Predator console retains a control stick similar to those found in manned aircraft. Its presence is a familiar token for pilots and supports an illusion of hands-on control. According to one Predator pilot, however, the control stick is pretty much a "morale stick"—it is not required since maneuvers can be directed via keyboard inputs.[74] It is a nonessential artifact of an older paradigm. Similarly, RPA consoles are not equipped with rudder pedals since such basic control inputs are managed via computer. In the RPA world, traditional stick-and-rudder inputs would actually undermine precise control due to a two-second lag-time imposed by distance and computer processing.[75] The pilot would have a delayed sense of what the aircraft is actually doing; she could not sense movement in any axis and would likely aggravate the situation with untimely inputs. RPA pilots thus do not exercise the manual skills prized in traditional aviation.

Another key element differentiating manned and unmanned aircraft involves the general architecture of the drone "cockpit." Predator and Reaper crews sit in a Ground Control Station; Global Hawk crews call it a Mission Control Element.[76] Each setup resembles a control room rather than a cockpit. The crew can use phones or computer chatrooms to engage a spectrum of outsiders, such as meteorologists, intelligence analysts, weapon experts, lawyers, flight planners, technicians, supervisors, and various personnel in the combat zone.[77] This, in turn, means that RPA crews are more accessible to external players and significantly less independent than crews of manned aircraft. Supervisors and other interested parties may intervene via phone, computer, or potentially just walk in to the control room. Other crewmembers show up for shift changes or perhaps to provide additional assistance if the scenario demands it. Unlike manned aircraft, RPAs are more of an open system than a closed one since they enable various outside experts to enter the control loop alongside the pilot to influence tactical details and overall outcomes.[78] The relatively permeable control architecture thus disperses expertise beyond the pilot, the traditional locus of authority and control.

All of this indicates that unmanned aviation demands a basic reevaluation of the pilot's role. It is a different paradigm that conflicts with the established pilot-centric worldview of flight. Unsurprisingly, the ongoing effort to gain intellectual mastery of unmanned aviation creates new questions about pilot identity.

Originally, the pilots of Global Hawk, Predator, and Reaper were mostly dragooned from the manned aircraft community. Few pilots volunteered for RPA duty. As the demand for drone flights rapidly increased, the Air Force ordered more and more pilots to the unmanned realm. A senior officer at Creech Air Force Base, the installation near Las Vegas that serves as the primary hub of Predator and Reaper operations, admitted in 2009 that Creech had become "a leper colony." Many pilots stationed there felt trapped and shunned as their careers rotted away. A general perception emerged that pilots were exiled to Creech if they were just average, undistinguished performers in their home community of manned aircraft.[79] The Air Force's reluctance to release RPA pilots for career-enhancing opportunities such as a premium staff tour or war college assignment exacerbated this stigma.[80] Yet this is ironic since the notable success of Predator and Reaper in the Middle East owed, among other things, to the cutting-edge work of these "leper colony" pilots.[81]

To fill the increasing demand for RPA operators and simultaneously reduce the number of experienced pilots ordered to Creech, in 2009 the Air Force began sending newly minted inexperienced pilots straight to RPAs with the promise they would then fly manned aircraft a few years later.[82] It also permitted combat system operators, nonpilots extensively trained as navigators or air battle managers, to fly RPAs. To some degree this shift to brand new pilots and combat system operators reflected the institution's pragmatic efforts to reduce the number of midcareer pilots sent to RPAs. But it also illustrated that the skills necessary to operate RPAs in a dynamic and complex combat environment were substantially different from those for manned aircraft and warranted a different type of pilot. Accordingly, in 2010 a new Air Force career field was established and given its own Specialty Code: 18X, "RPA Pilot."[83]

Unlike bomber, cargo, fighter, or tanker pilots, all designated by an 11X Specialty Code, aviators in the new 18X RPA career field do not attend the Air Force's yearlong $557,000 undergraduate pilot training program.[84] Instead, RPA pilots attend a six-month undergraduate RPA training program at the relative bargain of $65,000.[85] It includes thirty-nine hours of initial flight training at the 1st Flying Training Squadron in Pueblo, Colorado, where officers seeking to become RPA pilots, combat system operators, or manned aircraft pilots obtain their first exposure to basic flight procedures and gain experience in fundamental aviation skills.[86] RPA students then migrate to Randolph Air Force Base for an RPA instrument qualification course that includes forty-nine hours of simulator sorties and then a fundamentals course emphasizing a month of classroom training in sensors, weapons, communications, and other elements of modern aviation's combat architecture.[87] They are then poised to integrate into operational RPA squadrons for three months of orientation before becoming fully qualified for missions taking place over various continents.

This emerging species of drone pilots means that in the RPA community pilots of manned aircraft work alongside and compete for promotion with pilots trained to fly only unmanned aircraft. In 2013, for example, 42 percent of pilots flying RPAs were officers who earned their traditional pilot wings in undergraduate pilot training and were slated to return to a manned aircraft after their RPA tour. Another 40 percent were traditional pilots permanently recategorized into the RPA career field who would never return to manned aircraft. The remaining 18 percent were "18X" RPA-only pilots.[88] The Air Force plans to achieve 90 percent "18X" RPA-only pilots and recently doubled its annual production from 182 to 384, as it released more pilots back to manned aircraft.[89] Thus, the pilots who will gain intellectual mastery over an increasingly powerful regime in military aviation will be ones who never went to traditional pilot training.

The Air Force's December 2015 decision to make RPA-only pilots eligible for the same $25,000 annual aviation bonus pay awarded manned aircraft pilots indicates the value of this new type of pilot to institutional leaders.[90] Although this monetary incentive was motivated by the need to stem the tide of all types of pilots leaving uniformed service, it also helps bridge the gap and establish parity between drone pilots and traditional pilots in Air Force aviation. The Air Force is also aware of the need to improve promotion rates and career development opportunities for RPA pilots in order to make them long-term, influential members of the military aviation community.[91] Nevertheless, especially given the issue of physical risks accepted by manned aircraft pilots, some might ask, "Are drone pilots *real pilots*?"

This question is unsurprising since the paradigmatic difference between manned and unmanned flight confuses the professional identity of pilots. Yet flight in general has been redefined—and pilot function transformed—by displacing human control with machine control to overcome human limits, a key argument of this book. Thus, asking if "drone pilots are *real pilots*" is somewhat irrelevant since what it means to be a real pilot has changed dramatically in aviation's evolution, regardless if one actually sits in an aircraft or not. Such a question echoes the old paradigm of leather-jacketed, white-scarfed pilots and ignores the realities of modern flight. It forgets the advice to "get over your paradigms."

Conclusion

Airpower enthusiasts past and present have stood tiptoe on a ladder of technoscientific achievements to peer at the possible future of flight. The publication of *Toward New Horizons* in 1945, a document that contemplated such developments as "remote-controlled fighter and bomber forces," resulted from the collaboration of General Hap Arnold and Dr. Theodore von Kármán as they looked toward the future of aerial warfare. With the benefit of decades of subsequent advances in aeronautics, biology, physics, and computer processing, in 2010 one of General Arnold's successors charted the future of human augmentation and autonomous technology in a new volume, *Technology Horizons*. The Air Force extended this vision of aviation's evolution with a 2015 report whose title acknowledged its technoscientific heritage and anticipated even more changes for the pilot-aircraft complex: *Autonomous Horizons*.

Scientific and technological developments have permitted an ongoing reimagination of flight. This includes an evolving complementarity between pilot and aircraft where either may override the other's authority and control under certain conditions. It also involves the establishment of what the secretary of the Navy described as the "new normal," the popularity of unmanned and increasingly autonomous systems. Whatever the new normal may turn out to be, the traditional view of what it means to be a pilot is obscured by the growing cloud of drones on the horizon. As in the past, pilots will be challenged not only in their purpose and identity but also in the need to gain intellectual mastery over new forms of flight.

Despite the obvious differences between manned and unmanned aircraft, their analogous use of computer-based control has illustrated the fundamental similarities between the pilots who operate them. The pilots of most modern aircraft, manned and unmanned alike, are integrated into an elaborate cybernetic system where control is typically effected through highly automated systems reliant on computerized processing of vast amounts of information.[92] Traditional pilots and drone pilots must excel in a similar variety of tasks: maintain cognitive awareness of what the automatic systems are doing; deftly manage disparate internal and external subsystems to optimize aircraft function; remain alert to environmental factors such as weather, obstacles, and enemy threats; and communicate effectively with other aircrew members and supporting agents, including air traffic control, ground-based commanders, and command and control aircraft. Thus, in many ways, what it means to be a modern pilot transcends the manned and unmanned worldviews of flight. So rather than ponder if drone pilots are real pilots, let us take the advice of the Predator squadron commander to "get over your paradigms" and consider instead the future of pilots and how they may remain influential in aviation, manned or unmanned.

The Past and Future of Pilots

Modern aviation is not a reflection of pilot skill but a result of pilot failings. This is not a criticism of pilots but a recognition of the fundamental changes in the pilot–aircraft relationship that occurred before the Jet Age, Space Age, and Computer Age. As machine performance increased faster than human performance in the early decades of flight the problems with pilots became starkly evident. Their physical and cognitive limitations stunted the potential of manned flight, and overcoming these problems redefined flight by opening the cockpit to new technological interventions. Pilots ceded their traditional hands-on, seat-of-the-pants control to superior forms of mechanical control. They abandoned their own senses and instincts to rely on an instrument-based version of reality, cockpit automation, and external inputs. While flight surgeons and engineers created innovative ways to mitigate human frailties with machine enhancements, airpower enthusiasts—influential leaders often personally familiar with the skills, thrills, and perils of flight—sought to push the envelope of aircraft performance even if it required significant changes in pilot behavior. By the end of World War Two, pilots served as human servomechanisms alongside sophisticated control technologies in a complex cybernetic system, and this phenomenon informs the roles and expectations of modern and future pilots.

The problems with pilots that were identified in the early decades of flight can still prevail, as the story of Lieutenant Colonel Kevin Henry demonstrates. The introduction described how he narrowly escaped disaster flying his U-2 on a high-altitude reconnaissance mission in March 2006. The environmental extremes of flight gave him a severe case of decompression sickness (DCS) and disrupted his neurological functions to the point where he could not utilize the aircraft's instruments or autopilot as he struggled toward home. He was advised repeatedly to eject and barely managed to land. Yet this near-death experience was not quite the final chapter in his aviation career.

Undeterred by his harrowing DCS episode, Lieutenant Colonel Henry sought to fly the U-2 once again. The neurological tests performed in the hours and days after he was pulled semiconscious from his aircraft were followed by more tests in the weeks and months after he returned home to Beale Air Force Base, California. After numerous medical evaluations, flight surgeons deemed him fully capable of returning to the cockpit. He was fit to fly, or so he thought. On 15 September 2006 Lieutenant Colonel Henry was once again cruising above 65,000 feet. Sealed in his pressure suit and breathing pure oxygen, he sat and observed as the autopilot flew a preprogrammed course on a routine training mission in the skies of northern California. After two hours at high altitude and while reprogramming the navigation computer, decompression sickness struck once again.

Kevin Henry recalls, "I had a DCS brain hit—dizziness and vertigo—felt like I was going to pass out (scared the heck out of me, actually)."[1] He remembers drifting into a dreamlike state, free-falling through the atmosphere. As with his previous episode in March he had difficulty reading his navigational instruments, but he recognized the familiar geography near Beale Air Force Base and landed before his symptoms worsened. He knew this landing would be his last one in the U-2. After Lieutenant Colonel Henry's second DCS episode, a team of flight surgeons, neurologists, and psychiatrists at the US Air Force School of Aerospace Medicine, the successor to the School for Flight Surgeons established in 1918, determined that as a pilot he was a machine beyond repair. Among other injuries, physicians discovered that the "blue cones were dead" in his retinas, thus creating a permanent form of colorblindness. He could no longer function reliably in flight.[2]

Kevin Henry's ordeal is instructive as a modern example of the classic hazards of flight and how humans have sought to overcome them with new technologies of life support and automation. Yet it also raises questions about the future of pilots. As he descended for his final landing, he shared the sky above Beale Air Force Base with another type of high-altitude aircraft whose pilots were utterly immune to the perils of decompression sickness: the unmanned Global Hawk. U-2 pilots will quickly and correctly point out that the Global Hawk lacks many of the capabilities of the U-2; nevertheless, the fact that Kevin Henry's world of high-altitude reconnaissance was now shared by aircraft governed primarily by intricate and partially autonomous software illustrates how military leaders have exploited alternatives to traditional piloted aircraft. After a summary of the essential changes in the pilot–aircraft relationship we will return to what all of this means for the possible roles of humans in the future of flight.

How Flight Was Redefined

Detlev Bronk, a biophysicist and president of the National Academy of Sciences from 1949 to 1953, noted in 1948 that "the progress of aviation has depended on the combined efforts of physiologists, physicists, and engineers. . . . Time and again physicists and engineers have developed machines that the flier has been unable to use."[3] This observation illustrated that the transformation of flight involved more than progress in aerodynamics and aeronautical engineering; it also concerned advances in the life sciences, as illustrated in chapters 1 and 2. Flight surgeons resembled aeronautical experts: both enabled improvements in aviation, and both conducted their own versions of engineering science. As engineers of the body, flight surgeons tested the parameters of human function and applied this information to the design of equipment that enhanced aircrew survival.

Flight surgeons also contributed to the scientific foundation of the air force by nurturing a wide variety of disciplines in aviation medicine. They described their work as a new hybrid that included laboratory research, preventive medicine, internal medicine, physiology, otolaryngology, neuropsychiatry, psychology, anthropometry, and ophthalmology.[4] Aviation medicine also included the engineering of life-support equipment, and as airmen struggled against various environmental hazards, flight surgeons gained greater influence in aircraft design. In 1935, for example, the Air Corps established the Physiological Research Unit (PRU) at Wright Field, the hub of research and development in Army aviation. Under the leadership of Captain Harry G. Armstrong, the PRU enabled aviation medicine to "exert its influence on the design and operation of aircraft."[5] Armstrong worked with physiologists, test pilots, and aeronautical engineers and applied his knowledge of the body in flight to the development of pressurized aircraft. Other flight surgeons researched the physiology of oxygen starvation and crafted essential oxygen equipment and pressure-breathing technology. They also employed special devices such as the altitude chamber and human centrifuge to develop an empirical understanding of pilot physiology. Many of these life scientists complemented their search for knowledge with pragmatic solutions that shaped military aviation years before the Second World War.

The engineering science practiced by flight surgeons and aviation engineers enhanced the potential of manned aircraft, and it also triggered a new paradigm of flight within a decade after the Great War. The inability of pilots to remain oriented at night or in poor weather posed a significant limitation to aviation potential, yet this problem was resolved through a combination of science and engineering. As described in chapter 3, the 1926 collaboration between a flight surgeon and an experienced pilot resulted in a new understanding of man in flight. Major David Myers and Major William Ocker combined scientific knowledge of the human vestibular system with the use of a gyroscopically powered turn indicator, discovering that pilots could not fly in low-visibility conditions unless they focused on special flight instruments instead of their own senses. Instrument flight thus emerged from the interaction of science and engineering, and it fundamentally altered both the role of the pilot and the capability of aviation.

Human limitations also spurred new attitudes toward the necessity of instrumentation and automation. In its redefined form, flight no longer required pilots to rely solely on their own senses; instead, they used machines to interpret the environment for them when human senses became utterly unreliable. Chapter 3 tells how pilots learned to operate within a new paradigm of flight in which they developed an "instrument consciousness" and controlled their aircraft based on instrument data rather than their personal interpretation or "feel."[6] Flight took on an entirely new aspect, moreover, when pilots relinquished control to automatic devices. Flight became an activity in which the pilot surrendered more and more of his judgment and control to instrument and automation technology that could fly the aircraft with greater precision than humanly possible.

During the interwar years and World War Two, flight was also redefined in terms of system design, the subject of chapter 4. The effort to integrate human and machine components for maximum performance of the overall system laid the foundation of human factors engineering and portended what Norbert Wiener later called "the new science of cybernetics."[7] As the cybernetic system of flight grew more complex and automation equipment improved, different tasks were parceled out to the component—either human or machine—that could execute them more efficiently. To improve system performance, the air force utilized rigorous training methods to standardize pilot functions, and checklists were employed to guide a pilot's thoughts and discipline his machine-like actions. Chapter 5 reveals how in some cases, such as unmanned B-17s, flight no longer required a direct human presence, and the pilot could be displaced from the aircraft. Flight still relied on human judgment, but such judgment could be rendered through remote-control devices in a cybernetic system governed by sophisticated means of feedback and self-regulation.

All of these technoscientific efforts in the early decades of flight enhanced the military potential of aircraft while altering the role of the human component and magnifying the influence of physicians, engineers, and practically minded aviation leaders outside the cockpit. These predecessors of the modern Air Force reimagined the pilot's role in an instrumented, automated, and increasingly elaborate cybernetic system to improve aviation's efficacy and expand institutional power. Their efforts, combined with the constant adaptations of pilots, redefined flight.

Master or Servant?

The tension between pilot-based and machine-based control is fundamental to aviation's evolution. Increasingly effective means of automation enhanced the safety and efficiency of flight and expanded its commercial and military potential, but this also relegated pilots to new roles inconsistent with their professional ethos of hands-on control. Pilots nevertheless sought to maintain authority over a swiftly changing airman–aircraft complex.

In the early decades of flight pilots frequently recalibrated their role in order to retain control and preserve their status, and others have investigated how such efforts shaped the technology of flight. Historian David Mindell, for example, observes that pilots needed to develop "different kinds of intuition and skill" to deal with technological advances such as instrument-based flight and automatic flight control systems.[8] Generally, pilots would accept the erosion of their customary form of control and adopt new skills if they believed doing so would expand the aircraft's capabilities and enhance their professional standing.[9] Mindell also argues that the pilot–aircraft relationship was characterized by a technical philosophy that kept pilots, and eventually astronauts, "in the loop" and able to exercise some form of decisive agency, especially during critical phases of flight.[10] This is an important element in explaining the relationship between aviators, engineers, and their machines. Indeed, the effort of pilots to still play some sort of hands-on role, even where machine control was demonstrably superior, and exert final authority over the operation of the system permeates the evolution of aviation technology. Yet this book argues that something more is going on: in many ways pilots became servants to their technology rather than masters of it. This is evidenced by their redefinition as machines and managers.

An underlying feature of the redefinition of flight was the portrayal of the human in machine terms. Pilots can be viewed in many respects as human servomechanisms, "meat sprockets," in an elaborate system that is only as strong as its weakest component.[11] Flight surgeons, clinical pathologists of flight who probed the limits of human function, treated pilots as machines highly susceptible to failure and whose biomechanical functions limited system potential. They integrated pilots with special life-support technologies so the human body would remain functional while exposed to the environmental extremes of flight. Pilots submitted themselves to procedures and equipment that warmed their body, squeezed blood into their brain, and forced oxygen into their lungs. One 1943 Army Air Forces training manual instructed pilots that from ground level to 10,000 feet "You are a **flyer**," but when climbing above ten thousand feet toward treacherously thin and frigid air "You are a redesigned **machine**."[12] Maintaining the function of the overall system required precise knowledge of the human machine and its treatment as a standardized component that could be efficiently integrated with the aircraft.

Another aspect of characterizing pilots as machines involves the need to standardize their capabilities and discipline their actions. In their physical bodies and actions, they needed to be reliable as well as replaceable by other human machines. Pilots had to internalize the need to submit to superior instruments and develop the rote skills associated with instrument-based flight as their organic senses were augmented or replaced by cockpit technology. They relied on exogenous inputs such as radio beacons or ground-based radar control to guide their actions. Pilots became components—sometimes the weakest component—in a complex cybernetic system. As the pilot's role evolved other agents such as life scientists and engineers achieved greater status in aviation. This phenomenon was encouraged by airpower enthusiasts who sought to increase the power of the aviation system as a whole.

In their pursuit of air dominance, airpower enthusiasts, as noted in the introduction, required human operators to adapt to machines and even become like machines themselves. Pilots transformed into checklist-driven standardized servomechanisms embedded within a cybernetic system and instructed accordingly that above 10,000 feet "You are a redesigned machine." This may not have suited the preferences of individual pilots who did not see themselves in mechanistic terms, yet they had to submit themselves to new roles to reliably serve the larger interests of the system.

This fundamental change in what pilots do was necessary to improve the reliability and potential of aircraft for commercial and military purposes. Where they could not be machine-like and when their human capabilities fell short, pilots gave way to devices that were more effective and could optimize the performance of the system as a whole. This required them to serve as managers overseeing other machine components rather than exerting direct control.

Modern pilots serve a managerial role as they monitor various sensors and automation devices that perform the basic duties of flight. This was evident in the aircrew operating the ill-fated Air France Flight 447. Conditioned to a largely hands-off role by an aviation system far safer and more reliable when aircraft are controlled by automatic systems instead of human reflexes, they lacked the manual flying skills necessary in an emergency to take physical control of their Airbus at high altitude and stop its long plummet into the Atlantic. Similar to airline pilots, military pilots devote much of their cognitive energy to maintaining awareness of what the automatic flight control systems are doing and what they might do next. Both types of pilots are also aware that in some aircraft the automatic systems limit the options and range of inputs the pilot may make anyway. The phenomenon of sitting as a safety observer and managing the normal function of the aircraft's redundant control systems challenges the customary framework that portrays the interaction between airman and aircraft as a relationship engineered to favor pilot dominance. Instead, as Norton Schwartz, the top general in the Air Force, acknowledged in 2010, "We must now reconsider the relationship of man and woman, machine, and air."[13]

Some might argue that this reconsideration of the pilot–aircraft relationship signals out-of-control technology that has become adversarial and oppressive. Historian of technology Lewis Mumford warned in 1963 of an "authoritarian technics" that stripped humans of their agency in order to empower a machine system with superior "energy, speed, or automation."[14] His views are echoed by a later observer's claim that technology may "subordinate human capabilities to its own impersonal, destructive logics of rationality and domination" and demote the human to a "cog in the machine."[15] Mumford also argued that an elite corps of technocratic managers seeks to expand system control even if it diminishes human control. They are compelled to enhance the performance of a cybernetic, highly automated system by reducing its reliance on an undisciplined and variant human component that Mumford described as "resistant, sometimes actively disobedient servo-mechanisms, still human enough to harbor purposes that do not always coincide with those of the system."[16]

Mumford may have been tempted to view the actions of airpower enthusiasts and the transformation of the pilot's role through this lens. After all, airpower enthusiasts were driven by the effort to expand institutional influence, resulting in leaders such as General Hap Arnold and his modern equivalent General Norton Schwartz openly encouraging the pursuit of superior systems that do not rely on the traditional role of pilots. They were animated by the notion that advancing the power of the machine equates to advancing the power of their institution. Accordingly, we see evidence of authoritarian technics in microcosm in the pilot–aircraft relationship: the pilot's role transformed from sole physical master to a sometimes redundant component. Automation usurped pilot control, and pilots now behave more like standardized mechanical cogs in a complex and powerful system increasingly resistant to the actions of "disobedient" individuals.[17] One surmises this trespass on individual influence and the erosion of pilot agency would have alarmed Mumford and affirmed his fears.

Yet in his portrayal of authoritarian technics as "system-centered immensely powerful" technology that devalues human agency, Mumford also notes this is not an inevitable outcome if humans remain vigilant and limit the authority of technological systems.[18] This lack of inevitability makes room for human agency and undermines claims that technology has slipped the leash of human control. The evolution of the pilot–aircraft relationship may display some alarming similarities to authoritarian technics that eclipse human control, but it is not an inescapable victory of machine over man. Human agency persists.

Like other technology, aviation evolves within a social context shaped by cultural, political, and economic factors.[19] Mumford's concerns are offset by the fact that pilots ceded control not to a mindless authoritarian system but to experts outside the cockpit eager to advance their own ideas and influence. Physicians, engineers, and airpower enthusiasts sought to maximize the utility and reliability of the aviation system as a whole to fulfill their own ambitions and institutional goals. If and when necessary, they would use pilots. This phenomenon underlies a change in the nature of flight. Although humans are still essentially in control—they engineer technology, conduct research, write algorithms, fund systems, set goals, make rules, and adopt innovations—pilots have increasingly relinquished their customary role.[20] Flight thus remains a story of human mastery, but not necessarily *pilot* mastery. So what might the future hold for pilots?

New Futures, New Problems

The traditional image of the leather-jacketed, white-scarfed pilot may persist in some part. After all, Air Force and Navy pilots still sport leather jackets (scarves: optional). But the modern reality of who a pilot is or what he or she does clashes with such iconic portrayals. Instead, the airman–aircraft relationship has evolved into a construct of human–machine complementarity, where the machine may exert a dominant role that limits pilot inputs or overrides them altogether. We see this, for example, in sensors that detect and evaluate environmental threats long before the pilot can—an F-35 pilot peers through "the jet's eyeballs to see the world as the jet sees the world"—and also in autonomous systems that limit or even seize control to prevent pilot mishandling.[21] This is further evident in what may become the "new normal" of military aviation, remotely piloted aircraft. Modern drones displace the pilot altogether from the machine so he can join sensor operators, intelligence officers, and other nonpilots to better serve the system from a control station on the ground. As we consider what these developments might mean for the future of pilots, let us first assess why the skies aren't already thronged with unmanned aircraft.

Unmanned aircraft were neither inevitable nor surprising. The direction of aviation's evolution is governed inherently by human decisions, societal priorities, and institutional and economic dynamics. It is not predetermined by technoscientific advances to arrive at a particular type of aircraft, manned or unmanned. However, the early development and now the growing flock of unmanned aircraft make sense given the political and institutional imperatives to exploit the command of the air in the most effective manner possible and at the least risk to human life. These imperatives were powerful enough for the Air Force, ironically a pilot-centric institution led by pilots, to cultivate unmanned aircraft that decenter the pilot and undermine traditional pilot culture. Yet why is the emphasis on establishing a whole new category of pilot to operate remotely piloted aircraft only a recent phenomenon?

The development of unmanned aircraft has been part of military aviation for nearly a century. The US Navy conducted developmental tests of the Curtiss-Sperry Flying Bomb in 1917–18, and the US Army Signal Corps' Kettering Bug—a hybrid of wings, gyroscope, cash register, and player piano—lurched from its launch rail in 1918.[22] The coupling of automatic pilots to radio- and radar-based control resulted in far more complex technologies, such as the unmanned B-17 and B-24 bombers of Project Aphrodite in the Second World War. In the second half of the twentieth century the Air Force and Navy continued to develop a variety of unmanned aircraft, namely for a niche reconnaissance role and frequently under the cloak of secret programs affiliated with the Central Intelligence Agency and the National Reconnaissance Office.[23] Only with the recent advent of Predator drones over the Balkans and then over the Middle East did unmanned systems come into the public eye and gain traction within broader military aviation.

Although the Air Force and to some degree the Navy devoted considerable resources to developing unmanned systems over the past six decades, these systems were upstaged by vastly greater emphasis on manned flight and long-range missiles.[24] Budget resources not claimed by manned bombers, fighters, transports, and similar aircraft tended to flow toward ICBMs and cruise missiles rather than unmanned aircraft. Unmanned power projection in the Cold War era was thus emphasized mostly in the form of nuclear-tipped missiles, weapons of intense interest and interservice rivalry since they were deemed more important to the West's strategic goals than unmanned aircraft.

Additional reasons unmanned aircraft garnered less attention until recently include a comparative lack of interservice rivalry over them, the managerial impediments associated with developing secret projects outside of the mainstream procurement process, and the reluctance to spend money on questionably reliable systems whose purpose was generally satisfied by manned systems.[25] In addition, perhaps the most compelling reason for the delayed expansion of unmanned systems into the broader ecosystem of military aviation was a lack of technological sophistication and reliability.[26] The exponential advances in computing power in the age of Moore's Law, however, reduced that obstacle and permitted far-reaching prospects for civilian and military aviation.

Integral to the future of manned and unmanned aircraft are improvements in autonomous technology enabled by advances in processing power, computerized sensors, and artificial intelligence. We have seen how past technoscientific advances such as those associated with aviation medicine exerted enormous influence on the design and operation of aircraft as well as what was expected of pilots. Radar serves as another illustration of a new ability reshaping the potential of an entire system. For example, in 1945 Theodore von Kármán noted radar's ability to create new opportunities in bombing, gunnery, navigation, aircraft detection, ground-based control, and the "provision of information and controls to relieve the overburdened pilot." He also recognized how this new capability could greatly extend "the range, power, capabilities, and accuracy of human vision."[27] Like radar, autonomous technology spawns a ubiquitous array of possibilities and a reimagining of the potential of aircraft and the role of humans.

Chapters 6 and 7 assessed numerous developments tied to improvements in automatic and autonomous systems. Commercial and military pilots now rely heavily on various forms of automatic control and electronic sensors. Fly-by-wire systems ensure aircraft remain within safe parameters by limiting the degree of human input. Automatic flight control systems link with navigation and sensor data to relieve humans from their customary tasks of aircraft control and navigation. Automatic ground collision avoidance systems in the latest fighters are constantly poised to seize control when they compute that a crash is imminent, and it is likely the Air Force will soon adopt similar autonomous systems that take over and execute complex maneuvers to avoid air-to-air collisions as well. Yet there are more aspects to autonomy's promises for military aviation.

Pilots strapped into the world's most expensive fighter aircraft, the F-35, rely on an active electronically scanned array radar that automatically assesses targets at all ranges and provides a "protective sphere around the aircraft for missile warning, navigation support, and night operations."[28] According to Northrup-Grumman technicians, this capability needs no physical input from the pilot who acts only as "an administrator." The technicians also note engineers "make the jet easy to fly so the pilot can focus on mission management."[29] Another aspect of this latest aircraft is its fusion of information about the operating environment. Its six day- and night-vision cameras plus its onboard sensors and radar stream data into the pilot's helmet; much of these data are highly processed and prioritized before they ever reach the pilot's senses.[30] Indeed, in some scenarios the aircraft may decide to take action before the pilot has a chance to react.

The modern cockpit generates an evolving complementarity between pilot and aircraft. Both are brittle in different ways: automatic pilots and autonomous systems historically lack the situational awareness and versatile judgment possessed by pilots; pilots lack the precision control and computational speed of machines. Each partner compensates for the other's weaknesses to optimize function of the cybernetic system as a whole. The trend in machine intelligence, however, indicates autonomous technology is becoming far less brittle and far more flexible. This portends a further shift in the human–machine balance as aviation potential expands.

In 2010 the chief of staff of the Air Force predicted a reevaluation of the pilot–aircraft relationship. Subsequent Air Force policy characterizes this relationship in rather flexible terms. For example, the Air Force chief scientist testified to Congress in 2015 that the Air Force Research Laboratory's vision for autonomy is "intelligent machines seamlessly integrated with humans, maximizing mission performance in complex and contested environments."[31] Note the vision's careful word choice. Machines will integrate with *humans*, not *pilots*. The chief scientist assured Congress that the "approach is to keep the *airman* [not pilot] at the center of the system." (The only time he used the word pilot was elsewhere in his testimony when explicitly referring to manned aircraft.) The generic term *airman*, however, is preferred when discussing human-autonomy teaming. This suggests a human role but not necessarily a pilot role. Moreover, the claim that machine intelligence and autonomy permit future operations in combat scenarios exceeding human limitations indicates that the future human–autonomy partnership, although "seamless," may not be equal.[32]

In addition to transforming the pilot–aircraft relationship, autonomy's advance will likely fuel new capabilities for RPAs just as radar opened up new vistas for manned aircraft. One possibility involves controlling multiple aircraft simultaneously. The Air Force chief of staff declared in 2009 that the idea of one airman flying one aircraft was now "a very Neanderthal way of operating."[33] He was reflecting on the concept of multiaircraft control (MAC) where one RPA pilot operates her own formation of several aircraft. Despite the top general's enthusiasm, MAC failed to become an operational reality largely due to insufficient technological capabilities.[34] The Neanderthals had not sufficiently evolved. Recent advances in artificial intelligence, however, may make this evolutionary step attainable. In 2015 an experimental form of artificial intelligence that can sift through volumes of data for relevant variables in a process deemed "fuzzy logic" demonstrated impressive abilities to operate multiple aircraft in dynamic and complex computer simulations. Nicknamed "ALPHA," it even bested a senior aviator in dogfighting scenarios, shooting him down in each simulated engagement.[35]

ALPHA's ability to simultaneously control numerous aircraft and coordinate their adaptive responses to enemy action will likely not escape the attention of airpower enthusiasts. It suggests the ability to operate multiple drones that interact autonomously to meet an objective assigned by controllers on the ground. Such advances may also warrant a blended formation of manned and unmanned aircraft where the pilot of the manned aircraft assigns his unmanned wingmen various offensive and defensive tasks without having to directly manage their individual maneuvers. The Air Force Test Pilot School demonstrated this capability in 2015 when an unmanned F-16 successfully flew in formation with its manned counterpart, conducted dynamic aerial procedures, and performed key combat-related maneuvers.[36] The algorithm-controlled unmanned F-16 passed a more advanced test in 2017 when it autonomously reacted to emerging threats while conducting ground strikes in a mock combat environment.[37] One wonders what this means for the future of military aviation as these leaps in technology catalyze new thinking. Yet new thinking spawns new problems, a theme in aviation and technology in general.

Perhaps Captain Chesley "Sully" Sullenberger, the pilot who glided his engine-out Airbus safely onto the Hudson River in 2009, said it best: improvements in machine control technology "change the nature of the errors that are made."[38] Sullenberger's observation echoes chapter 4's discussion of how aviation's advances in the interwar years and World War Two fostered new types of pilot error and designer error. In 1928, for example, the National Advisory Committee for Aeronautics created various categories such as "error of judgment," "poor technique," and "carelessness or negligence" to help understand why pilots flew good airplanes into the ground.[39] As the cybernetic system of flight became increasingly complex, it generated new errors in system design that aggravated pilot misjudgment. The US Army Air Forces Aero Medical Psychology Branch illustrated this concept by emphasizing the importance of designing cockpit technology that pilots could integrate with intuitively and efficiently in order to remain aware of what was happening. Their work anticipated the problems familiar to Sullenberger and other modern pilots—namely, the lurking dangers associated with loss of cognitive control.

One form of the loss of cognitive control, as described in chapter 6, is automation surprise. By relieving pilots of many traditional duties of hands-on control, automatic flight control systems tend to generate complacency and misperception. On occasion the aircraft will do something, or fail to do something, the pilot anticipated. Such matters concern commercial and military aviation and contributed to a 2013 FAA report highlighting how overreliance on automatic systems dulls pilot skill and awareness.[40] Some NASA researchers noted that automation surprises reflect the "increasing autonomy and authority of machine agents," especially when they act "on their own" without specific direction from the pilot.[41] This phenomenon contributes to the broader problem of the loss of cognitive control.

The investigation into the Air France Flight 447 tragedy determined that the pilots experienced a "total loss of cognitive control of the situation."[42] The Airbus 330's cybernetic system that relied on pilot–aircraft complementarity, especially in an unusual flight scenario such as the one encountered over the Atlantic that evening, broke down as the pilots failed to correctly intervene due to their misunderstanding of what their cockpit systems told them. Essentially, the problem was that they didn't anticipate the possibility of these events. As autonomy advances, therefore, how will pilots retain cognizance of an increasingly complex system designed to think and act faster than humans? How will they keep up?

This returns us to Norbert Wiener, the father of cybernetics. After spending two decades establishing cybernetics, a term he used to describe "the entire field of control and communication theory,"[43] Wiener reflected on its consequences in a 1960 essay. Rejecting the common opinion that machines will remain simple and controllable, entirely limited and predictable in their operation, he warned that they might "transcend some of the limitations of their designers" and operate at speeds and levels of analysis far in excess of human capability.[44] This was a prescient observation in 1960 given the leaps in computational power in the subsequent five decades. System designers and operators, therefore, may be unable to understand machine intentions and operations. This astutely predicted how modern pilots are subject to automation surprise and loss of cognitive control. It also applies to the problem of trust calibration described in chapters 6 and 7 and how an enduring problem with pilots includes how to get them to trust and submit to machine autonomy in the appropriate circumstances.

Norbert Wiener also warned of "disastrous results . . . in the real world wherever two agencies essentially foreign to each other are coupled in the attempt to achieve a common purpose."[45] He was referring to humans and machines and the problem that both "operate on two distinct time scales." This informs and complicates the 2015 *Autonomous Horizons* report that portrays pilots and autonomous technology as a partnership or team. The advances in machine autonomy lend credence to Wiener's prediction that pilots won't be able to keep up, so what will be their role in this partnership? Will the history of autonomy be the story of pilots losing it and aircraft (manned and unmanned) gaining it? Wiener ended his essay with an exhortation: "We must always exert the full strength of our imagination to examine where the full use of our new modalities may lead us."[46] As flight evolves, therefore, human imagination and intellectual mastery still matter.

Conclusion

People with the title of "pilot" will likely be around for a long time. They are not history yet. But the main point is that the pilot's traditional role in the machine—one based on native wit and physical mastery—was fundamentally transformed even before humans entered the supersonic realm of jets and sped toward the modern age of machine intelligence. In the decades before the Second World War, communities of physicians, engineers, and airpower enthusiasts developed means to either strengthen or replace the weak link in order to advance the promise of technology. In this froth of innovation, the pilot evolved from sole master of his machine into a human servomechanism surrounded by a cumbersome life-support apparatus, disciplined to trust instruments over instinct, trained to submit to superior automation devices, subjected to external control, and in some cases evicted from the aircraft altogether. All of this deposed the pilot as lone sovereign and empowered others to fight for command of the air.

This redefinition of the pilot's role anticipated a cybernetic age where computers and automation disrupted old roles and created new ones for vast numbers of people. It also permitted five-star General Henry "Hap" Arnold, the US Army Air Force's top Airman in 1945, to envision a future when "we won't have any men in a bomber." Apparently, those most interested in exploiting the full potential of flying machines knew about the inherent problems with pilots for a long time. What they did about it redefined the human–machine relationship in aviation and opened the door even wider to a modern era in which human agency integrates and competes with machine intelligence.[47]

The airman–aircraft relationship has evolved into a new construct of human–machine complementarity where the machine may exert a dominant role that limits human inputs or overrides them altogether. We see this, for example, in sensors that evaluate environmental threats long before the pilot can and also in autonomous systems that limit or even seize control to prevent pilot mishandling. This is further evident in what may become the "new normal" of aviation, remotely piloted aircraft, and their displacement of the pilot from the machine so he or she can join sensor operators, intelligence officers, and other nonpilots to better serve the system from a control station on the ground.

The growing momentum of unmanned aircraft perpetuates the original appeal of airpower: the ability to project power cheaply and effectively in the third dimension. This vision impelled airpower enthusiasts to seek various ways to optimize flight and overcome human and machine limitations. The problems with pilots have thus changed, but the presence of pilots will likely endure in some form. They may continue to serve as a locus of human agency in a complex system, even as their physical and cognitive roles are reimagined and flight is redefined to suit human and institutional ambitions.

Coda

Lieutenant Colonel Kevin Henry is a brave pilot and a fortunate soul. He survived two episodes where nitrogen bubbles, in protest of low atmospheric pressure, ran riot in his joints and brain. The effects were so severe in the first episode that he could hardly summon the effort to actuate his airplane's controls. Nor could he read the flight instruments or engage the automation devices that are now essential to command of the air. Yet he survived, though now forbidden from the cockpit by flight surgeons. Countless others who sought mastery of the airman–aircraft complex were not as fortunate. Indeed, the hard lessons and technological advances that redefined flight in the early twentieth century are too often made apparent in modern tragedies.

John F. Kennedy Jr., his wife, Carolyn, and her sister, Lauren, were flying from Essex County Airport in New Jersey to Martha's Vineyard on the evening of 16 July 1999 for a wedding the next day. John was at the controls of his personal airplane, a single-engine Piper Saratoga fully equipped with blind-flight instrumentation and an automatic pilot. He had earned his private pilot's license about two years earlier and flown the route back and forth to Martha's Vineyard numerous times. They took off at 8:40 PM for what should have been a routine flight. The weather forecast looked good, a pleasant summer evening with no storms or fog. Other pilots flying near Martha's Vineyard, though, observed significant haze and no discernible horizon. One pilot wondered if there was a power failure at Martha's Vineyard because he couldn't see the lights. Still, everything seemed normal as John, Carolyn, and Lauren flew eastward into the night.

According to investigators from the National Transportation Safety Board (NTSB), the trouble started nearly an hour after takeoff.[1] Just thirty-four miles west of Martha's Vineyard, John's aircraft entered a descending right turn, pointing away from the destination. A minute later he stopped the descent at about 2,500 feet above the Atlantic and rolled out of the turn. He then began a left turn back toward the Vineyard. During this turn, he started to descend once again, this time at a fairly steep rate of 900 feet per minute. He rolled out of the turn for a few seconds but continued descending. Then, according to radar data, the aircraft entered a steep right turn. The airspeed increased rapidly, and they plummeted toward the water at the excessive rate of 4,700 feet per-minute. At this point, John had less than thirty seconds to recognize the danger and recover to level flight. It wasn't enough time.

US Navy divers found the wreckage four days later. The Office of the Chief Medical Examiner of Massachusetts decreed that all three "died from multiple injuries."[2] NTSB investigators stated the probable cause: "The pilot's failure to maintain control of the airplane during a descent over water at night, which was a result of spatial disorientation."[3] As described in chapter 3, spatial disorientation in flight results from the human inability to remain oriented without visual reference to a fixed horizon. As darkness and haze made it impossible to discern ocean from sky, John's ability to use his own senses to determine his proper relationship to the Earth below became undone. As his aircraft spiraled toward the water in a steep right turn, it is probable that he thought he was turning the other way and perhaps even climbing. He had entered a situation where flight was not possible for humans, at least not without elaborate technological enhancements that present an artificial interpretation of reality to refute the lies told by human senses. Kennedy's aircraft had functional blind-flight instruments on board, but he lacked the "instrument consciousness" and experience necessary for their proper use.[4] In the conditions that night, he could not fly the airplane without them, yet he did not realize this in time. He lost his way, and then lost everything.

This second Kennedy aviation tragedy illustrates why physicians, engineers, and airpower enthusiasts redefined flight in the first half of the twentieth century. Humans could not exploit the utility of aircraft without surrendering their own senses and judgment to superior machines. The pressure to expand aviation capabilities required an increasingly elaborate cybernetic system capable of achieving higher altitudes, longer ranges, and greater precision in nearly all weather conditions. Pilots became decentered in this system. Instead of the central control element, they were one of many components, often the weakest component. Their tasks were replaced by superior mechanical forms, even to the point of physical removal from the aircraft in order to achieve various military goals.

In 1944 a different Kennedy tragedy occurred. Lieutenant Joseph Kennedy Jr. attempted to parachute from an unmanned B-24 Liberator bomber flown via remote control, but he perished in a blinding explosion. In 1999 his nephew and two other family members lost their lives when the pilot failed to disengage his own senses and rely instead on either automatic control or flight instruments. Both tragedies—generations apart—highlight how flight has been redefined and transformed into a cybernetic affair where human presence could be a liability. As the airman–aircraft relationship continues to transform, both tragedies also remind us that the story of an evolving symbiosis with technology is, in the end, a human story.

Notes

INTRODUCTION: Pilot Problems and Machine Promises

1. Kevin M. Henry, interview by author, Beale Air Force Base, CA, 20 June 2006. Some aspects of Kevin Henry's story were also related to the author via telephone and email correspondence between July 2006 and October 2008. For additional information on Kevin Henry's story and episodes of decompression sickness involving other U-2 pilots, see Mark Betancourt, "Killer at 70,000 Feet: The Occupational Hazards of Flying the U-2," *Air & Space Magazine*, May 2012, http://www.airspacemag.com/military-aviation/killer-at-70000-feet-117615369/; James A. Marshall, "The Greatest Flying Story Never Told," *Combat Edge*, May 2010; and Roger Aylworth, "U-2 Pilot Miraculously Survives Harrowing Flight," *Oroville Mercury-Register*, 5 November 2006, http://www.orovillemr.com/article/ZZ/20061106/NEWS/611069986.

2. U-2 pilots remain sealed within their pressure suit. They drink by inserting a straw through a one-way valve. Similarly, they use straws to eat a paste-like food, some of it fortified with caffeine. During long sorties they may also eat an amphetamine-laced "go-gel." For a history of the U-2 program and a description of pilot activities, see Chris Pocock, *The U-2 Spy Plane: Toward the Unknown* (Atglen, PA: Schiffer Military History, 2000).

3. For a fuller exposition of DCS sequelae and a clinical diagnosis of this incident, see S. L. Jersey et al., "Severe Neurological Decompression Sickness in a U-2 Pilot," *Aviation, Space, and Environmental Medicine* 81, no. 1 (January 2010): 64–68.

4. The U-2 squadron conducting operations in this region was hosted by a US ally. The actual location is referred to here as "Home Base."

5. David L. Russell, interview by author, Beale Air Force Base, CA, 20 June 2006. See also Aylworth, "U-2 Pilot Miraculously Survives Harrowing Flight."

6. After the conclusion of this entire incident, Russell queried a flight surgeon about Henry's visual impairment and learned it was similar to the symptoms of a stroke; in this case, a nitrogen bubble rather than a blood clot impaired blood flow to ocular functions. David Russell, personal correspondence, 6 April 2017.

7. Kevin Whitelaw, "A U-2 Pilot's Dangerous Landing," *U.S. News and World Report*, 20 September 2007, http://www.usnews.com/news/national/articles/2007/09/20/a-u-2-pilots-dangerous-landing.

8. Jersey et al., "Severe Neurological Decompression Sickness in a U-2 Pilot," 65.

9. Ibid. The flight surgeon's on-scene diagnosis was "severe DCS with neurological symptoms and incipient cardiovascular collapse."

10. Kevin M. Henry, interview by author, Beale Air Force Base, CA, 15 May 2007. See also Jersey et al., "Severe Neurological Decompression Sickness in a U-2 Pilot," 64–68. Lieutenant Colonel Russell recalls the flight surgeons discussing the numerous hyperbaric chamber treatments, or "dives," required by Lieutenant Colonel Henry. After the first dive, the flight surgeons gained confidence he would survive; the second dive affirmed his normal functions would largely return, and a third dive convinced them he would be at least 90 percent normal. After the fourth dive he felt back at 100 percent. Russell, personal correspondence, 6 April 2017.

11. Kevin Henry's U-2 exhibited a left-hand rolling tendency, which aggravated his spiraling descent. Lieutenant Colonel Russell repeatedly ordered, "Raise your left, Kevin" and thus likely minimized the descent rate. Russell, personal correspondence, 6 April 2017.

12. Major Henry H. Arnold, "The Performance of Future Airplanes," 6 May 1925. RG 18, entry 22, box 502, National Archives, College Park, MD. By 1938, Major General Arnold was the chief of the Army Air Corps and ultimately achieved five-star rank as the nation's top airman.

13. Ibid., 4.

14. Historian David Mindell makes a similar observation in his analysis of the relationship between humans and machines in spaceflight. He observed that pilots learned "to trust instruments more than themselves, making the transition from 'natural' pilots who flew by sense, to 'mechanical' ones who flew by rules and indicators." See David Mindell, *Digital Apollo: Human and Machine in Spaceflight* (Cambridge, MA: MIT Press, 2008), 24.

15. A 1991 NASA Technical Memorandum made a similar observation: "In the early days of aviation, the pilot set forth unaided, with only human perceptual capabilities to provide necessary information. It was soon discovered that these were insufficient, and aircraft sensors and instruments were developed to augment the limited human capabilities." Charles E. Billings, "Human-Centered Aircraft Automation: A Concept and Guidelines," Ames Research Center, CA, 1991.

16. Thomas P. Hughes, *American Genesis: A Century of Invention and Technological Enthusiasm, 1870–1979* (Chicago: University of Chicago Press, 1989), 3. For additional examples of technological enthusiasm, see David Mindell, *Our Robots, Ourselves* (New York: Viking Press, 2015), 6–7.

17. Thomas P. Hughes, *Human-Built World: How to Think about Technology and Culture* (Chicago: University of Chicago Press, 2005), 53.

18. The term *heroic pilot* appears in other investigations of aviation and captures the social stereotype of aviators, particularly lone pilots enduring and conquering the duress of combat. Historian David Mindell employs this term to illustrate how some viewed autopilot technology as tolling the demise of "the era of heroic pilots." David Mindell, *Between Human and Machine: Feedback, Control, and Computing before Cybernetics* (Baltimore: Johns Hopkins University Press, 2002), 80.

19. Sydney Shalett, "Arnold Reveals Secret Weapons, Bomber Surpassing All Others," *New York Times*, 18 August 1945.

20. Norton Schwartz, interview by Tom Bowman, National Public Radio, 26 August 2008, http://www.npr.org/templates/story/story.php?storyId=93981179.

CHAPTER ONE: The Pathology of Flight

1. T. S. Rippon and E. G. Manuel, *Lancet*, 28 September 1918, 411–15.

2. Brigadier General Henry H. Arnold, speech to the Western Aviation Planning Conference, 23 September 1937, *U.S. Air Corps U-Stencils* 25, no. U-1203 (September 1937): 1–27. In 1924 the commandant of the US Army's School of Aviation Medicine argued that "there is no such thing as an aviator's disease," but aviators "are prone to develop certain diseases due to the character of their work and environment." See L. H. Bauer, "Diseases Due to Aviation, Including the Effects of Altitude and the Care of the Flyer," *Air Service Information Circular* 5, no. 462 (1 May 1924): 11.

3. For information on performance improvements in aviation between World War One and Two, see John G. Paulisick et al., *R&D Contributions to Aviation Progress (RADCAP), Summary Report*, vol. 1 (Springfield, VA: National Technical Information Service, 1972). This report was sponsored by the Aeronautical Systems Division, Air Force Systems Command, Wright-

Patterson Air Force Base, OH. See also Major Henry H. Arnold, "The Performance of Future Airplanes," 6 May 1925. RG 18, entry 22, box 502, National Archives, College Park, MD.

4. Walter M. Boothby, W. Randolph Lovelace II, and Otis O. Benson, "High Altitude and Its Effect on the Human Body," *Air Corps News Letter* 24, no. 2 (15 January 1941): 1.

5. Elliot White Springs, *War Birds: Diary of an Unknown Aviator* (New York: George H. Doran, 1926), 232, quoted in Douglas H. Robinson, M.D., *The Dangerous Sky: A History of Aviation Medicine* (Henley-on-Thames, Oxfordshire: G. T. Foulis, 1973), 76.

6. The "napping" German pilot's entry into a slow roll and his lack of defensive maneuvering suggest that he had somehow become incapacitated or unable to fight. Hypoxia was a likely culprit.

7. Bauer, "Diseases Due to Aviation," 11.

8. "How to Avoid Death from Oxygen Failure," (1945). Call number 248.4211-25, iris number 00183270, AFHRA, Maxwell Air Force Base, AL.

9. "Anoxia Deaths, Report No. 8." Call number 248.4211-25, iris number 00163270, AFHRA. This is one of seventeen hypoxia fatality reports from operational units forwarded to the air surgeon.

10. Harry G. Armstrong, "The Influence of Aviation Medicine on Aircraft Design and Operation," *Journal of the Aeronautical Sciences* 5, no. 5 (March 1938): 197.

11. Robert Boyle, "New Pneumatic Experiments about Respiration," *Philosophical Transactions of the Royal Society of London* 5, no. 63 (1670): 2044.

12. Ibid.

13. Ebbe Curtis Hoff, *A Bibliographical Sourcebook of Compressed Air, Diving and Submarine Medicine* (Washington, DC: Navy Department Bureau of Medicine and Surgery, 1948), 39.

14. In 1943 the Army Air Forces School of Aviation Medicine defined decompression sickness as "the physiological effect of reduced barometric pressure, independent of the effects of hypoxia." *Flight Surgeon's Handbook*, 2nd ed. (Randolph Field, TX: School of Aviation Medicine, Army Air Forces, 1943), 91. For more on decompression sickness, see John L. Phillips, M.D., *The Bends: Compressed Air in the History of Science, Diving, and Engineering* (New Haven, CT: Yale University Press, 1998), and Hurley L. Motley et al., "Studies on Bends," *Journal of Aviation Medicine* 16, no. 4 (August 1945): 210–24.

15. The US Army Air Forces Air Surgeon Office claimed in 1945 that "among the many medical problems associated with military aviation none is more exclusively limited to this field than the effect of centrifugal [acceleration] force on the flyer." See "Notes from the Air Surgeon's Office," *Journal of Aviation Medicine* 16, no. 1 (February 1945): 45.

16. In 1938 Dr. Harry G. Armstrong, Captain, US Army Medical Corps, conducted a seminal investigation into acceleration forces using aircraft and the human centrifuge. Armstrong's paper, co-authored with J. W. Heim, describes the symptoms experienced by pilots and lab subjects and traces the subject literature back to French experiments on dogs in 1917. See "The Effect of Acceleration on the Living Organism," *Journal of Aviation Medicine* 9 (December 1938): 199–233. Armstrong also summarized research in acceleration forces in his landmark text, *The Principles and Practice of Aviation Medicine* (Baltimore: Williams and Wilkins, 1939).

17. Levy Hathaway, "Flying Puts Strains upon the Human Body," *New York Times*, 11 November 1928.

18. Luke Christopher, "Upside Down and Tail First!" *Aeronautic Review* 7, no. 11 (November 1929): 18.

19. Robinson, *The Dangerous Sky*, 190. For an overview of the physiological and technical challenges of excessive G-forces, see John L. Brown and Marian Lechner, "Acceleration and Human Performance: A Survey of Research," *Journal of Aviation Medicine* 27, no. 1 (February 1956):

45. See also Harry G. Armstrong, ed., *Aerospace Medicine* (Baltimore: Williams and Wilkins, 1961).

20. J. L. Birley, "Temperament and Service Flying," *Reports of the Air Medical Investigation Committee*, 25 September 1918. RG 18, box 115, National Archives, College Park, MD. Birley was "Medical Officer in Charge R.A.F. in the Field."

21. Ibid., 46.

22. "The Air Medical Service and the Flight Surgeon," *Air Service Information Circular* 1, no. 4 (24 February 1920): 4.

23. Donald W. Hastings et al., "Psychiatric Experiences of the Eighth Air Force, First Year of Combat," August 1944. Box 1349, USAF Museum, Wright-Patterson Air Force Base, OH. This is a report on psychiatric episodes among flyers in the Eighth Air Force from July 1942 to July 1943.

24. H. I. Brock, "The Flying Man Is Etched in Court: An Admirable Picture of the Aviator Stands Out in the Testimony Given at the Mitchell Trial," *New York Times*, 6 December 1925.

25. Extract from Memorandum for Committee on Military Affairs, US House of Representatives, 25 February 1920. Call number 145.93-5, AFHRA. The death rate due solely to accident was almost forty-nine times higher for flying officers than nonflying officers.

26. Armstrong, "The Influence of Aviation Medicine," 194.

27. C. L. Beaven, "A Chronological History of Aviation Medicine," School of Aviation Medicine (Randolph Field, TX, 1939), 8, document number 168.7082-623, AFHRA. See also Timothy F. Kirn, "Army, Navy, Air Force, Reserve Flight Surgeons Share Aerial Working Conditions of 'Patients,'" *JAMA* 261, no. 19 (19 May 1989): 2761.

28. Birley, "Temperament and Service Flying."

29. Beaven, "A Chronological History of Aviation Medicine," 8.

30. Ibid., 4–7. Beaven's chronology contains a copy of the 2 February 1912 memo. The memo also describes how candidates in the US Navy were required to "leap directly up, striking the buttocks with both heels at the same time" as well as hop throughout the room.

31. Benjamin D. Foulois, "Military Aviation and Aeronautics," *Journal of the Military Service Institution of the Unites States* 52, no. 181 (January–February 1913): 115. Foulois became the chief of the Air Service of the American Expeditionary Air Forces in 1918, and he eventually ascended to the post of chief of the Army Air Corps in 1931.

32. "Notes on Administration," School of Aviation Medicine (Randolph Field, TX: Randolph Field Printing Office, 1938), 46. Call number 168.7082-612, AFHRA. For a useful summary of the origins of aviation medicine in the US Army, see Mae M. Link and Hubert A. Coleman, *Medical Support: Army Air Forces in World War II* (Washington, DC: Office of the Surgeon General, USAF, 1955), 6–24, and Adrianne Noe, "Medical Principle and Aeronautical Practice: American Aviation Medicine to World War II" (Ph.D. diss., University of Delaware, 1989).

33. W. H. Wilmer, "The Early Development of Aviation Medicine in the United States," *Military Surgeon* 77, no. 3 (September 1935): 115–35.

34. L. H. Bauer and W. M. MacLake, "The Air Medical Service and the Flight Surgeon," *Air Service Information Circular* 1, no. 4 (24 February 1920): 6. See also Philip W. Andrews, "The First Half Century of the School of Aerospace Medicine: January 1918–January 1962," 2. Call number 168.7082-93, AFHRA.

35. Noe, "Medical Principle and Aeronautical Practice," 304. Adrianne Noe uses the term *gatekeeper* to describe flight surgeons in her research on aviation medicine.

36. Henry Horn, "Acid Test of the Airman," *New York Times*, 20 April 1919.

37. W. A. Scruton, "Examination of Applicants for Aviation Service, US Army—Disqualifying Factors in Fifteen Hundred Cases—Some Observations of Past-Pointing after Rotation," *Annals*

of Otology, Rhinology, and Laryngology 27, no. 2 (June 1918): 528–33. This article explained the mallet and pistol tests as products of exaggeration and rumor.

38. For a more in-depth analysis of how flight medicine related to early twentieth-century developments in aviation and medicine, see Noe, "Medical Principle and Aeronautical Practice."

39. James H. Cassedy, *Medicine in America* (Baltimore: Johns Hopkins University Press, 1991), 76–86. See also Jeffrey P. Baker, *The Machine in the Nursery* (Baltimore: Johns Hopkins University Press, 1996), 105.

40. For an excellent analysis of how technology such as blood and urine chemistry or X-ray imagery transformed medicine in the first quarter of the twentieth century, see Joel D. Howell, *Technology in the Hospital: Transforming Patient Care in the Early Twentieth Century* (Baltimore: Johns Hopkins University Press, 1995).

41. Baker, *The Machine in the Nursery*, and Cassedy, *Medicine in America*.

42. W. Bruce Fye, *American Cardiology: The History of a Specialty and Its College* (Baltimore: Johns Hopkins University Press, 1996), 18–19.

43. Kenton Kroker, "Washouts: Electroencephalography, Epilepsy and Emotions in the Selection of American Aviators during the Second World War," in *Instrumental in War: Science, Research, and Instruments between Knowledge and the World*, ed. Steven A. Walton (Leiden: Brill Academic Publishers, 2005), 301–38.

44. Detlev W. Bronk et al., eds., *Advances in Military Medicine Made by American Investigators Working under the Sponsorship of the Committee on Medical Research*, vol. 1 (Boston: Little, Brown, 1948), 325–27.

45. Noe, "Medical Principle and Aeronautical Practice," 2, 10. Noe argued that flight surgeons exploited new devices such as human centrifuges and altitude chambers to establish their special role and bolster their authority in the Army and the medical profession. Similarly, historian Joel Howell examined how medical technology was incorporated into different hospitals and how its use was socially constructed. His analysis suggested that flight medicine's imperative to appear relevant in the peacetime environment may have influenced the use of chamber and centrifuge technology.

46. Deborah J. Coon, "Standardizing the Subject: Experimental Psychologists, Introspection, and the Quest for a Technoscientific Ideal," *Technology and Culture* 34, no. 4 (October 1993): 768.

47. Ibid., 783.

48. Ibid., 764.

49. John B. Watson, "On Behaviorism, 1913," in *A Source Book in the History of Psychology*, ed. Richard J. Herrnstein and Edwin G. Boring (Cambridge, MA: Harvard University Press, 1965), 507.

50. "Report of the Medical Research Laboratory and School for Flight Surgeons for the Calendar Year 1920," *Air Service Information Circular* 3, no. 231 (20 May 1921): 6.

51. John Carson, "Army Alpha, Army Brass, and the Search for Army Intelligence," *Isis* 84, no. 2 (June 1993): 278–309.

52. "Report of the Medical Research Laboratory and School for Flight Surgeons for the Calendar Year 1920," 9.

53. Eugene G. Reinartz, "Aviation and the Doctor," *Texas Reports on Biology and Medicine* 3, no. 3 (Fall 1945): 389. Brigadier General Reinartz served as the commandant of the School of Aviation Medicine during World War Two.

54. Ibid., 393.

55. *Air Service Information Circular* 5, no. 403 (1 March 1923): 1–80.

56. See, for example, the *Journal of the Aeronautical Sciences* 10, no. 6 (June 1943). Similarly, in 1944 an article by three Army flight surgeons describing current medical activities in military aviation contrasted with the journal's usual fare of aeronautical techno-speak, such as "Study of Exhaust-Valve Design from Gas Flow Standpoint" and "Strain Energy Analysis of Incomplete Tension Field Web-Stiffener Combinations." See 11, no. 1 (January 1944).

57. Malcolm C. Grow, "Flying Status of Flight Surgeons," *Flight Surgeon Topics* 1, no. 4 (October 1937): 20.

58. F. C. Dockeray and S. Isaacs, "Psychological Research in Aviation in Italy, France, England, and the A.E.F.," *Air Service Information Circular* 3, no. 237 (15 July 1921): 26–37.

59. Rebecca Hancock Cameron, *Training to Fly: Military Flight Training, 1907–1945* (Washington, DC: Air Force History and Museums Program, 1999), 124–25.

60. W. F. Craven and J. L. Cate, eds., *The Army Air Forces in World War II*, vol. 7 (Chicago: University of Chicago Press, 1947), 366.

61. "Why Should the Flight Surgeon Fly?" *Flight Surgeon Topics* 1, no. 3 (July 1937): 22.

62. H. H. Arnold, "Flying Status for Flight Surgeons," *U.S. Air Corps U-Stencils* 26, no. U-1279 (March 1939): 4.

63. Bosley Crowther, "'Dive Bomber,' a Colorful Film about Flight Surgeons, at the Strand—'Wild Geese Calling,' at the Roxy," *New York Times*, 30 August 1941, 10.

64. Stanley J. Reiser, *Medicine and the Reign of Technology* (New York: Cambridge University Press, 1978), 140.

CHAPTER TWO: Engineering the Human Machine

1. For a summary of the evolving views of natural philosophers, physicians, and scientists regarding the mechanical nature of life, see David F. Channell, *The Vital Machine: A Study of Technology and Organic Life* (New York: Oxford University Press, 1991).

2. Ann Thomson, ed., *La Mettrie: Machine Man and Other Writings* (Cambridge: Cambridge University Press, 1996).

3. Claude Bernard, *An Introduction to the Study of Experimental Medicine* (New York: Macmillan, 1927).

4. Joel D. Howell, *Technology in the Hospital: Transforming Patient Care in the Early Twentieth Century* (Baltimore: Johns Hopkins University Press, 1995), 208.

5. Theodore C. Lyster, "The Aviation Service of the Medical Department of the Army," *Annals of Otology, Rhinology, and Laryngology* 27, no. 3 (September 1918): 854. Brigadier General Lyster was the first chief surgeon of the Army Signal Corps Aviation Section and a patriarch of aviation medicine.

6. Thomas P. Hughes, *Networks of Power* (Baltimore: Johns Hopkins University Press, 1983).

7. Thomas P. Hughes, "Convergent Themes in the History of Science, Medicine, and Technology," *Technology and Culture* 22, no. 3 (July 1981): 554.

8. For excellent analyses of airpower strategy in the interwar years and World War Two, see Michael S. Sherry, *The Rise of American Airpower* (New Haven, CT: Yale University Press, 1987), and Tami Davis Biddle, *Rhetoric and Reality in Air Warfare* (Princeton, NJ: Princeton University Press, 2002).

9. Hughes, *Networks*, 15.

10. Ibid., 22, 14.

11. G. B. Obear, "A Note on Oxygen Supply for Aviators," *Air Service Information Circular* 1, no. 3 (1920): 44, and G. B. Obear, "A Note on the Low-Pressure Chamber Installed in the School of Aviation Medicine," *Air Service Information Circular* 5, no. 403 (1923): 75. Doctor Obear served as head of the Medical Research Laboratory's Department of Physics and Engineering.

12. Edward C. Schneider, Brenton R. Lutz, and Harold W. Gregg, "Compensatory Reactions to Low Oxygen," *Air Service Information Circular* 1, no. 99 (1920): 62.

13. Harry G. Armstrong, "Anoxia in Aviation," *Journal of Aviation Medicine* 9 (June 1938): 84.

14. Henry Horn (Lieutenant Colonel, US Army), "Acid Test of the Airman," *New York Times*, 20 April 1919.

15. Leon T. LeWald and Guy H. Turrell, "The Aviator's Heart—Roentgen Ray Studies under Conditions Simulating High Altitudes," *Air Service Information Circular* 1, no. 3 (1920): 3. Wald was a roentgenologist, and Turrell was chief of the Cardiovascular Department at the Medical Research Laboratory. Both physicians held the rank of major in the US Army Medical Corps.

16. Obear, "A Note on the Low-Pressure Chamber," 75.

17. "Report on Altitude Flight of Major R. W. Schroeder," Air Service Engineering Division, McCook Field, Dayton, OH. Small manuscript series 1349, box 1, Clark Special Collections Branch, USAFA Library, Colorado Springs, CO.

18. Detlev W. Bronk et al., eds., *Advances in Military Medicine Made by American Investigators Working under the Sponsorship of the Committee on Medical Research*, vol. 1 (Boston: Little, Brown, 1948), 213.

19. Apollo Soucek, "Navy Flier Attains 40,000 Feet to Set New Altitude Mark: Aviator Writes Own Story of His Battle to Breathe and See at Great Height," *New York Times*, 9 May 1929.

20. L. H. Bauer, "Diseases Due to Aviation, Including the Effects of Altitude and the Care of the Flyer," *Air Service Information Circular* 5, no. 462 (1924): 13. Bauer was the commandant of the School of Aviation Medicine.

21. A. P. Gagge, S. C. Allen, and J. P. Marbarger, "Pressure Breathing," *Journal of Aviation Medicine* 16, no. 1 (February 1945): 2. Pressure breathing should not be confused with the pressure-differential mechanism of the iron lung. For an explanation of how the iron lung functions, see David J. Rothman, *Beginnings Count: The Technological Imperative in American Health Care* (Oxford: Oxford University Press, 1997), 42–43.

22. A. Pharo Gagge, Colonel, USAF, informal interview, Medical Research Division, Office of the Surgeon General, Headquarters USAF, Washington, DC, 1 February 1951. RG 18, entry 22, box 60, National Archives, College Park, MD. See also A. Pharo Gagge, "The War Years at the Aeromedical Lab: Wright Field (1941–1946)," *Aviation, Space, and Environmental Medicine* 57, no. 10 (October 1986): A6–A10.

23. Charles A. Dempsey, *50 Years of Research on Man in Flight* (Wright-Patterson Air Force Base, OH: United States Air Force, 1985), 32.

24. James P. Henry, informal interview, Acceleration Unit, Aero Medical Laboratory, Wright-Patterson Air Force Base, OH, 8 September 1950. RG 18, entry 22, box 60, National Archives, College Park, MD. Dr. Henry served as a medical research consultant and chief of the Acceleration Unit.

25. John R. Poppen, "Recent Trends in Aviation Medicine," *Journal of Aviation Medicine* 12, no. 1 (December 1941): 55.

26. Hurley L. Motley et al., "Studies on Bends," *Journal of Aviation Medicine* 16, no. 4 (August 1945): 210, 232.

27. Ibid., 220.

28. Green Peyton, *50 Years of Aerospace Medicine: 1918–1968* (Brooks Air Force Base, TX: Air Force Systems Command Historical Publications Series No. 67-180, 1968), 123.

29. Robert J. Benford, *The Heritage of Aviation Medicine: An Annotated Directory of Early Artifacts* (Washington, DC: Aerospace Medical Association, 1979), 29.

30. Ibid.

31. Helen W. Schulz, *Case History of Pressure Suits* (Wright-Patterson Air Force Base, OH: Historical Office, Air Materiel Command, 1951), 3.

32. "Flights in B-17E Aircraft by Crew, Using Pressurized High Altitude Pilot Suits and Equipment," Army Air Forces Materiel Division Memorandum Report, 30 October 1942. RG 18, entry 22, box 60, National Archives, College Park, MD.

33. "Final Report on Tests of Pressurized Altitude Suits," Proof Department, Miscellaneous Section, Army Air Forces Proving Ground Command, Eglin Field, FL, 11 November 1942. RG 18, entry 22, box 60, National Archives, College Park, MD.

34. "Report of Travel, Aero Medical Laboratory: High Altitude Pressure Suit," Engineering Division, Army Air Forces Materiel Command. 5 October 1943. RG 18, entry 22, box 60, National Archives, College Park, MD.

35. Memorandum from Vannevar Bush to Major General Henry "Hap" Arnold, Chief of the Air Corps, 31 October 1940. RG 18, entry 22, box 60, National Archives, College Park, MD.

36. Memorandum from the Army Air Forces Materiel Division to Vannevar Bush, 2 December 1940. RG 18, entry 22, box 60, National Archives, College Park, MD.

37. Major Henry H. Arnold, "The Performance of Future Airplanes." 6 May 1925. RG 18, entry 22, box 502, National Archives, College Park, Md. Arnold provided no scientific basis for human limitations; rather, he just surmised that there would have to be some method for "protecting the pilot" at high speed, and pilots might not be able to "manipulate fighting equipment fast enough to be effective."

38. Douglas H. Robinson observed that "the accuracy of dive bombing appealed to naval aviators, who were realizing since the Mitchell experiment that level bombers flying at high altitude stood little chance of hitting a vessel maneuvering at high speed." See Douglas H. Robinson, M.D., *The Dangerous Sky: A History of Aviation Medicine* (Henley-on-Thames, Oxfordshire: G. T. Foulis, 1973), 113–14.

39. For an overview of the physiological and technical challenges of excessive G-forces, see John L. Brown and Marian Lechner, "Acceleration and Human Performance: A Survey of Research," *Journal of Aviation Medicine* 27, no. 1 (February 1956): 45.

40. Edward W. Constant II, *The Origins of the Turbojet Revolution* (Baltimore: Johns Hopkins University Press, 1980), 15.

41. Ibid., 17. The concept of a pilot as presumptive anomaly also appeared in German aviation medicine research: "It is a manner of common sense to realize that any technical development enters a blind alley unless it includes an acceptance of the physiological efficiency of man as a determining factor when the effect of the functional entity aircraft-man is being calculated. With the increasing speed of high-efficiency aircraft, this factor assumes increasing importance. He who starts his calculation with this factor in mind will obtain the best results in the long run." See Otto Gauer, "The Physiological Effects of Prolonged Acceleration," *German Aviation Medicine, World War II*, vol. 2 (Washington, DC: US Government Printing Office, 1950), 583.

42. The Germans had already constructed their own centrifuge and began testing in 1934. For a detailed account of the German aviation medicine program, see Gauer, *German Aviation Medicine*. This text originates from original German documents recovered after World War Two and was prepared under the auspices of the surgeon general of the US Air Force. See also William G. Clark and Ralph L. Christy, "Use of the Human Centrifuge in the Indoctrination of a Navy Fighter Squadron in the Use of Antiblackout Equipment," *Journal of Aviation Medicine* 17, no. 5 (October 1946): 394.

43. Ninth Annual Report of the Chief, Materiel Division of the United States Army Air Corps, Fiscal Year 1935, box R1-357, History Office, Air Force Materiel Command (AFMC), Wright-Patterson Air Force Base, OH.

44. In 1938 Harry Armstrong and J. W. Heim conducted a landmark investigation into acceleration forces using aircraft and the human centrifuge. Their paper describes the symptoms experienced by pilots and lab subjects and traces the subject literature back to French experiments on dogs in 1917. See Harry G. Armstrong and J. W. Heim, "The Effect of Acceleration on the Living Organism," *Journal of Aviation Medicine* 9 (December 1938): 199–233.

45. "Notes from the Air Surgeon's Office," *Journal of Aviation Medicine* 16, no. 1 (February 1945): 45. See also Edward H. Lambert, M.D., "Effects of Positive Acceleration on Pilots in Flight, with a Comparison of the Responses of Pilots and Passengers in an Airplane and Subjects on a Human Centrifuge," *Journal of Aviation Medicine* 21, no. 3 (June 1950): 195–220.

46. For a groundbreaking investigation of how G-forces affected pilots, see Armstrong and Heim, "The Effect of Acceleration," 199–233.

47. For information on the US Navy's version of an early G-suit, see Clark and Christy, "Use of the Human Centrifuge," 394–98. Interestingly, Sir Frederick Banting, co-discoverer of insulin in 1921, led Canada's development of a centrifuge in 1940. On 21 February 1941 he boarded a bomber flight to England to share his work with English researchers, but he was killed when the plane crashed on takeoff. Rebecca J. Green et al., eds., *50 Years of Human Engineering: History and Cumulative Bibliography of the Fitts Human Engineering Division* (Wright-Patterson Air Force Base, OH: Armstrong Laboratory, Air Force Materiel Command, 1995), 167.

48. "Notes from the Air Surgeon's Office: The G Suit," *Journal of Aviation Medicine* 16, no. 1 (February 1945): 46.

49. Philip J. Dorman and Richard W. Lawton, "Effect on G Tolerance of Partial Supination Combined with the Anti-G Suit," *Journal of Aviation Medicine* 27, no. 6 (December 1956): 490.

50. "Notes from the Air Surgeon's Office," 45.

51. Dorman and Lawton, "Effect on G Tolerance," 490. See also Vincent M. Downey et al., "Effect of the Crouch Position on the Increase in Tolerance to Positive Acceleration Afforded by an Antiblackout Suit," *Journal of Aviation Medicine* 20, no. 5 (October 1949): 289–99.

52. Lewis Wood, "Says We Beat Nazis in Aero Research," *New York Times*, 28 July 1945. This article summarizes an interview with Colonel William Randolph Lovelace, the head of the Army Air Forces Aero Medical Laboratory (formerly known as the Physiological Research Laboratory), after he returned from a tour of Germany.

53. Paul-Werner Hozzel, Brigadier General (ret.), German Air Force, served in the Second World War as a Stuka pilot. In a 1978 interview, Hozzel described a typical dive-bombing run: "When passing the [bomb release] altitude in the dive, a loud and clear horn signal was sounded, warning the pilot to press the bomb releasing button on the control stick and to pull out the plane. By pressing the releasing button, we also automatically actuated the hydraulic recovery device which aided the pilot, under the heavy G-load encountered in steep dive recoveries, in pulling out of the dive." Hozzel also describes how another Stuka pilot, Hans-Ulrich Rudel, sunk a Soviet battleship "but as [Rudel] pulled out on that attack he fell unconscious; the aircraft steered itself away apparently." See *Conversations with a Stuka pilot* (Columbus, OH: Battelle, 1978).

54. Withstanding 13.5 G, for example, required a blood pressure of 400 mm Hg, nearly 300 points above the average level. This risked damaging blood vessels or rupturing delicate structures in the lungs. Earl H. Wood, "Contributions of Aeromedical Research to Flight and Biomedical Science," *Aviation, Space, and Environmental Medicine* 57, no. 10 (October 1986): A13.

55. German researchers evaluated the supine position in 1937 and the prone position in 1939. See Gauer, *German Aviation Medicine*, 583.

56. C. E. Kerr et al., "Note on Prone Position in Aircraft," Royal Air Force Flying Personnel Research Committee Report (date indeterminate, but declassified on 7 Mar 1958), 4. Box 1349, USAF Museum, Wright-Patterson Air Force Base, OH.

57. Memorandum Report on "Prone Position" from the Army Air Forces Aero Medical Laboratory, Wright Field, OH. 25 February 1944, 1. Box 1349, USAF Museum.

58. Ibid., 2.

59. Memorandum Report on "A Prone Position Bed for Pilots," from the US Air Force Aero Medical Laboratory, Wright-Patterson Air Force Base, OH. 25 June 1948, 10. Box 1349, USAF Museum.

60. Ibid., 25.

61. Bronk et al., *Advances in Military Medicine*, 211, 251.

62. Ibid. A study in 1946 investigated the effects of various intravenously injected drugs on decompression sickness. Theophyllin was tested to see if it could ward off decompression sickness; it failed. Narcotics such as codeine and Demerol proved effective in relieving decompression-sickness-related pain, but their side effects eroded aircrew efficiency. See O. L. Williams et al., "The Use of Drugs for the Prevention of Decompression Sickness," *Journal of Aviation Medicine* 17, no. 6 (December 1946): 602–5.

63. *Air Service Medical* (Washington, DC: War Department, Air Service Division of Military Aeronautics, 1919), 3, 12.

64. Ibid., 7 (emphasis in original).

65. Harry Armstrong, "Laboratory to Test High-Altitude Effect," *New York Times*, 19 April 1936. Armstrong made similar claims in 1938: "Some of the most pertinent information in aviation medicine is not known, or is not accepted, by the aviation industry because the aeronautical engineer is interested primarily in aircraft performance and not in the human element and since the two are naturally incompatible the latter is allowed to suffer at the expense of the former." Harry G. Armstrong, "The Influence of Aviation Medicine on Aircraft Design and Operation," *Journal of the Aeronautical Sciences* 5, no. 5 (March 1938): 193.

66. Irving B. Holley Jr., "Technology and Doctrine," in *Technology and the Air Force, a Retrospective Assessment*, ed. Jacob Neufeld (Washington, DC: Air Force History and Museums Program, 1997), 102–3.

67. According to Bronk, "Time and again physicists and engineers have developed machines that the flier has been unable to use; or the machines endangered the lives of those who used them. Then the progress of aviation again waited on the assistance of the biologists. Thus we have attained our present prowess in the air through the combined efforts of the physical scientist and the physiologist." Bronk et al., eds., *Advances in Military Medicine*, 207.

68. For a brief summary of the problems associated with extreme cold, see Otis O. Benson and Ernest A. Pinson, "The Problems Inherent in the Protection of Flying Personnel against the Temperature Extremes Encountered in Flight," *Journal of the Aeronautical Sciences* 9, no. 7 (May 1942): 252–54.

69. Walter G. Vincenti, *What Engineers Know and How They Know It: Analytical Studies from Aeronautical History* (Baltimore: Johns Hopkins University Press, 1990), 51, 88.

70. Holley, "Technology and Doctrine," 103.

71. Dempsey, *50 Years of Research*, xxvii.

72. For official summaries of research and development efforts at Wright Field, see *A Century of Growth: The Evolution of Wright-Patterson Air Force Base* (Wright-Patterson Air Force Base, OH: Aeronautical Systems Center, 1999), and Albert E. Misenko and Philip H. Pollock, *Engineering History, 1917–1978: McCook Field to the Aeronautical Systems Division*, 4th ed., (Wright-Patterson Air Force Base, OH: Air Force Systems Command Historical Publication, 1979).

73. Adrianne Noe, "Medical Principle and Aeronautical Practice: American Aviation Medicine to World War II" (Ph.D. diss., University of Delaware, 1989), 186. On his own initiative, Grow worked on improving cold-weather clothing and reducing carbon monoxide levels in the cockpit (185).

74. Harry G. Armstrong, "The Medical Aspects of the National Geographic Society–U.S. Army Air Corps Stratosphere Expedition of November 11, 1935," *Journal of Aviation Medicine* 7 (June 1936): 55–62.

75. Pierre Leglise, "Ascent to the Stratosphere of the 'Explorer,'" Air Corps Technical Report, no. 303, trans. Ursula E. Athenstadt, 4 February 1935. Call number 168.6005-4, iris number 00123596, AFHRA.

76. Harry G. Armstrong, "The Medical Problems of Sealed High-Altitude Aircraft Compartments," *Journal of Aviation Medicine* 7 (March 1936): 2–8.

77. Dempsey, *50 Years of Research*, xxix.

78. Handwritten note by Malcolm C. Grow. This note was undated but clipped to original Air Corps memoranda from the spring of 1935 regarding the establishment of the Physiological Research Unit. The note read: "In a conference with General Robins [and] Col Martin . . . 2 weeks ago it was planned to proceed with the construction and equipment of a physiological lab at Wright Field." RG 18, decimal file 702.3, box 2800, National Archives, College Park, MD.

79. Memorandum from Brigadier General A. W. Robins, Chief of Materiel Division, Air Corps, to Chief, Engineering Section, Materiel Division, 6 May 1935. RG 18, decimal file 702.3, box 2800, National Archives, College Park, MD.

80. Memorandum from Brigadier General Oscar Westover, Assistant Chief of the Air Corps, to the Chief, Materiel Division, 29 May 1935. RG 18, decimal file 702.3, box 2800, National Archives, College Park, MD.

81. Memorandum from Colonel Frank M. Kennedy, Air Corps, Executive to the Chief of the Air Corps, "Monthly Progress Report for May of the Director, Physiological Research Laboratory," 9 June 1938. RG 18, decimal file 702.3, box 2800, National Archives, College Park, MD.

82. This story comes from an article by the original pilot who retired as a brigadier general in the US Air Force Reserve. See Harold R. Harris, "The Flying Steel Tank," *Air Power Historian* 4, no. 4 (October 1957): 213–15.

83. Robinson, *The Dangerous Sky*, 132–33.

84. Harris, "The Flying Steel Tank," 215.

85. Armstrong, "The Medical Problems of Sealed High-Altitude Aircraft Compartments," 2.

86. A. L. Klein, "Methods of Cabin Supercharging and Their Necessary Control Systems," *Journal of the Aeronautical Sciences* 3, no. 1 (September 1935): 58–60.

87. For an excellent analysis of the important role of aviation medicine in the development of the XC-35, see Seymour L. Chapin, "An Active Interface between Medical Science and Aeronautical Technology: The Physiological Investigations for the XC-35," *History and Philosophy of the Life Sciences* 13 (1991): 235–48.

88. For a more detailed description of the Lockheed XC-35 Electra, see the Smithsonian National Air and Space Museum article, http://www.nasm.si.edu/research/aero/aircraft/lockheed_xc35.htm.

89. Chapin, "An Active Interface," 236–37. Chapin notes that this was "an excellent example of effective collaboration between academe and government in the promotion of a significant technological development."

90. The account of this first hangar test comes from Alfred H. Johnson, "Supercharging a Pressure Cabin Airplane," *Journal of the Aeronautical Sciences* 5, no. 5 (March 1938): 175–80.

91. Armstrong, "The Medical Problems of Sealed High-Altitude Aircraft Compartments," 7–8.

92. J. W. Heim, "Physiologic Considerations Governing High Altitude Flight," *Journal of the Aeronautical Sciences* 5, no. 5 (March 1938): 189–92. Heim was a civilian physiologist in the Physiological Research Unit. These tests applied to flights below 50,000 feet. Rapid decompression above 60,000 would be much more serious: the time of useful consciousness was a matter of seconds, and decompression sickness was an immediate threat.

93. "Test Flights of New Sub-Stratosphere Lockheed Electra Plane," War Department press release, 6 August 1937. RG 18, box 2780, National Archives, College Park, MD.

94. For an explanation and history of the Collier Trophy, see the National Aeronautic Association website, http://www.naa.aero/html/awards/index.cfm?cmsid=62.

95. Memorandum from Vannevar Bush to Major General Henry "Hap" Arnold, Chief of the Air Corps, 31 October 1940. RG 18, box 60, National Archives, College Park, MD.

96. Bronk et al., eds., *Advances in Military Medicine*, 211, 213, 217.

97. Ibid., 207.

98. *Flight Surgeon Topics* 1, no. 2 (February 1937): 1. This is a publication from the School of Aviation Medicine at Randolph Field, Texas. Doc. no. 168.7082-610, AFHRA.

99. Copy of radiogram report to Military Intelligence Division, 1 May 1941. RG 18, box 60, National Archives, College Park, MD.

100. Routing and Record Sheet from General Arnold to General Brett, Materiel Division, 3 May 1941. RG 18, box 60, National Archives, College Park, MD.

101. Mae Mills Link and Hubert A. Coleman, *Medical Support: Army Air Forces in World War II* (Washington, DC: Office of the Surgeon General, USAF, 1955), vii.

CHAPTER THREE: Flying Blind

1. For an excellent portrayal of the development of instrument flight technology in the four decades following World War One, see Erik M. Conway, *Blind Landings: Low-Visibility Operations in American Aviation, 1918–1959* (Baltimore: Johns Hopkins University Press, 2006).

2. Winston Churchill, *Thoughts and Adventures* (London: Odhams Press, 1947), 128.

3. Ibid.

4. Ibid., 132. Churchill referred to a civilian pilot, Gustav Hamel, who solved the problem of fatal spins and taught military pilots a variety of maneuvers. In 1914 Hamel set out across the English Channel and disappeared. In his dramatic flair, Churchill wrote, "He had flown off in the fading light, into the squalls and mists of the Channel, confident that there was no difficulty and no danger he could not surmount, and from that moment he vanished for ever from human ken" (132). It is quite possible that Hamel became disoriented in the weather and crashed.

5. "Notes on Physiology in Aviation Medicine," *War Department Technical Manual*, no. 8-310 (21 October 1940). Call number 168.7082-628, iris number 1023183, AFHRA, Maxwell Air Force Base, AL. See also William C. Ocker, "Instrument Flying to Combat Fog," *Scientific American* 142, no. 5 (December 1930): 430.

6. Mason M. Patrick, *The United States in the Air* (Garden City, NY: Doubleday, Doran, 1928), 145. This story may be apocryphal since things dropping out of an aviator's pocket generally indicate a negative-G condition that is not usually associated with spatial disorientation. Nevertheless, General Patrick was correct to contextualize this story in 1928 with the newly appreciated fact that "without the aid of instruments it is impossible to tell whether a plane is flying level, is following a straight course, or is even right side up" (145).

7. Elmer J. Rogers Jr., "Blind Flying," *U.S. Air Services* 19, no. 4 (April 1934): 13.

8. Ibid., 31.

9. Ibid., 13.

10. Albert F. Hegenberger, "The Importance of the Development of Aerial Navigation Instruments and Methods to the Air Service," 1 September 1923, call number 168.7280-2, iris number 01105913, AFHRA.

11. For Mitchell's own account of this incident, see William Mitchell, *Winged Defense* (New York: G. P. Putnam's Sons, 1925), 66–73.

12. Rebecca Hancock Cameron, *Training to Fly: Military Flight Training, 1907–1945* (Washington, DC: Air Force History and Museums Program, 1999), 265.

13. Established in 1935, the GHQ Air Force was the organizational component of the Army Air Corps in charge of all combat units. The commander of the GHQ Air Force reported to the chief of staff of the Army. Responsibility for training, supply, doctrine, and research and development remained under the chief of the Air Corps.

14. Memorandum to the Chief of the Air Corps, 27 March 1939. RG 18, box 3, National Archives, College Park, MD.

15. Letter from Colonel Ira C. Eaker to Lieutenant General Delos C. Emmons, 28 January 1941. RG 18, box 1, National Archives, College Park, MD. Notably, General Emmons had written to the chief of the Air Corps nearly eighteen months earlier with a similar warning: "The requirements of war will force pursuit pilots to fly on instruments and it is essential that they receive the necessary training." See Memorandum from General Delos C. Emmons to the Chief of the Air Corps, 14 July 1939, RG 18, box 3. Despite General Emmons's entreaty, instrument training for pursuit pilots failed to gain priority status by the time of Eaker's letter. This may have been due, in part, to the limited role of fighter aircraft in the air force's preferred strategy of using well-armed heavy bombers.

16. Letter from Lieutenant General Delos C. Emmons to Colonel Ira C. Eaker, 1 February 1941. RG 18, box 1, National Archives, College Park, MD.

17. Hegenberger, "The Importance of the Development," 3.

18. For a basic summary of nineteenth- and early twentieth-century investigations into the vestibular system, see the Nobel Prize lecture given by Róbert Bárány on 11 September 1916. This lecture can be found in *Nobel Lectures, Physiology or Medicine, 1901–1921* (Amsterdam: Elsevier, 1967); it is also available from http://nobelprize.org/nobel_prizes/medicine/laureates/1914/barany-lecture.html.

19. Determining how temperature changes in the ear canal caused physical sensations of movement required considerable insight on the part of Bárány. Deviations from body temperature altered the specific gravity of the vestibular fluid, causing it to move. This, in turn, tricked the brain into perceiving that the body was undergoing some sort of angular motion. By changing the head's position by ninety degrees, Bárány triggered different sensations of movement based only on temperature changes. Bárány described this as his *experimentum crucis*, perhaps as a nod to Claude Bernard's use of the term in his famous 1865 text, *An Introduction to the Study of Experimental Medicine*. For a fuller account, see his Nobel Prize lecture: http://nobelprize.org/nobel_prizes/medicine/laureates/1914/barany-lecture.html.

20. Isaac H. Jones, *Equilibrium and Vertigo* (Washington, DC: Government Printing Office, 1917), 5. This publication is a transcript of Dr. Jones's lecture to the Medical Officer's Training Camps concerning evaluation standards for Army aviators.

21. Carl J. Crane, "Part Played by the Air Corps in Connection with Instrument Flying," *U.S. Air Corps U-Stencils* 25, no. U-1189 (1937–38): 2.

22. Jones, *Equilibrium and Vertigo*, 5.

23. The US Army Surgeon General, "Notes on Eye, Ear, Nose, and Throat in Aviation Medicine," *War Department Technical Manual 8-300* (26 November 1940): 72. Box 1347, USAF Museum, Wright-Patterson Air Force Base, OH. This report claims that before 1926 researchers focused on "finding a way of producing immunity against vertigo by means of placing pilots in a freely movable, revolving apparatus . . . in the belief that constant repetition of motion would establish an immunity. . . . Nothing came of this except to establish the fact that pilots who could expertly handle an orientator were possessed of a high degree of muscle sense and a keen sense of perception."

24. Robert J. Hunter, "Cultivating the Balance Sense: A Prelude to Cloud Flying," *Air Service Information Circular* 1, no. 3 (15 March 1920): 79–80.

25. "Aviation Instruments, a Brief History," *American Aviation Historical Society Journal* 6 (Fall–Winter 1961): 169.

26. Benjamin Foulois, interview by Peter Hackus, NBC *Today* program, 13 September 1963. Manuscript Series 17, box 10, Clark Special Collections Branch, USAFA Library, Colorado Springs, CO.

27. Letter from Sir W. H. D. Acland to Lieutenant Commander B. G. Connell, USN, Wright Field, OH, 2 November 1935. File G-2, box 848, USAF Museum.

28. Isaac H. Jones, "Blind Flying," in *An Outline of Otolaryngology as Applied to Aviation Medicine*, ed. John M. Hargreaves (Randolph Field, TX: School of Aviation Medicine, 1939), 74. Call number 168.7082, iris number 1023179, AFHRA, Maxwell Air Force Base, AL.

29. Mortimer F. Bates, "Filling up the Instrument Panel," *Aviation* (August 1941): 192. Bates worked for the Sperry Gyroscope Company, an important manufacturer of early aircraft instruments. The Creagh-Osborne compass was also used by infantrymen and, later in the war, tank crews.

30. Howard C. Stark, *Instrument Flying Instruction Book* (Pawling, NY: Distributed by James Stark, 1934), 15. File G-2, box 849, USAF Museum.

31. In September 1917, the Signal Corps issued Specification No. 27031, mandating that aircraft possess an "airspeed indicator, altimeter, magnetic compass, air pressure gauge, oil pressure gauge, gasoline level gauge, radiator water thermometer, tachometer, and clock."

32. "The Use of Instruments in Air Work," *Instruction Manual, Instruments, Technical Notes* (Air Information Signal Corps, 1918), 3. File G-2, box 848, USAF Museum.

33. Office of the Director of Air Service, "Utility of Airplane Instruments," *War Department Document*, no. 912 (Washington, DC: Government Printing Office, 1920), 5. File G-3, box 848, USAF Museum.

34. William C. Ocker and Carl J. Crane, *Blind Flight in Theory and Practice* (San Antonio, TX: Naylor Printing, 1932), 9–10. For more information on Elmer Sperry, see Thomas P. Hughes, *Elmer Sperry: Inventor and Engineer* (Baltimore: Johns Hopkins University Press, 1971), and David Mindell, *Between Human and Machine: Feedback, Control, and Computing before Cybernetics* (Baltimore: Johns Hopkins University Press, 2002).

35. Ocker and Crane, *Blind Flight*, 10.

36. "The Earth Inductor Compass," *Scientific Monthly* 25, no. 1 (July 1927): 91–92. Notably, Charles Lindbergh made the earth inductor compass famous when it guided him to within a few miles of his target in western Ireland. Lindbergh's 26 May 1927 cablegram from Paris to the Pioneer Instrument Company in Brooklyn read: "All instruments functioned perfectly during entire flight. I was less than five miles off course on Irish coast due largely to Earth inductor compass." Robert J. Scheppler, "Aviation Instruments: A Brief History, Part I," *American Aviation Historical Society Journal* 6, no. 4 (Fall 1961): 177.

37. Memorandum from War Department, Air Service, Engineering Division, Office of the Chief of the Division, to the Chief of the Air Service, Washington, DC, 22 October 1925. RG 18, box 2780, National Archives, College Park, MD.

38. Albert F. Hegenberger, "The Importance of the Development of Aerial Navigation Instruments and Methods to the Air Service," 1 September 1923. Call number 168.7280-2, iris number 01105913, AFHRA.

39. This expedition started on 6 April 1924 with four planes, and three planes finished 171 days later. David A. Myers, "The Medical Contribution to the Development of Blind Flying," in *An Outline of Otolaryngology as Applied to Aviation Medicine*, ed. John M. Hargreaves (Randolph Field, TX: School of Aviation Medicine, 1939), 96. Call number 168.7082, iris number 1023179, AFHRA.

40. "Air Service Aircraft Instruments, Operation and Maintenance," *Technical Regulation 1440-50* (1925): 1. File G-2, box 848, USAF Museum.

41. Memorandum from War Department, Air Service, Engineering Division, Office of the Chief of the Division, to the Chief of the Air Service, Washington, DC, 11 March 1925. RG 18, box 2780, National Archives, College Park, MD.

42. Ocker and Crane, *Blind Flight*, 17–18.

43. Myers, "The Medical Contribution," 90.

44. Ibid., 91. See also Maurer Maurer, *Aviation in the U.S. Army, 1919–1939* (Washington, DC, Office of Air Force History, 1987), 275–76.

45. For this account and a brief biography of Ocker, see William I. Chivalette, "Sergeant William Charles Ocker: The Army's Third Enlisted Pilot," *The Airmen Heritage Series* (Suitland, MD: Airmen Memorial Museum, 1992), 2.

46. Ibid., 5.

47. For Myers's account of this episode, see Myers, "The Medical Contribution," 91.

48. Ibid., 91, 93.

49. Ibid., 91.

50. David Mindell, *Between Human and Machine: Feedback, Control, and Computing before Cybernetics* (Baltimore: Johns Hopkins University Press, 2002), 15. Also see Steven Shapin and Simon Schaffer, *Leviathan and the Air Pump: Hobbes, Boyle, and the Experimental Life* (Princeton, NJ: Princeton University Press, 1989).

51. Thomas S. Kuhn, *The Structure of Scientific Revolutions*, 3rd ed. (Chicago: University of Chicago Press, 1996). 180.

52. Ibid., 6.

53. Ibid., 90, 169.

54. William C. Ocker, "Instrument Flying to Combat Fog," *Scientific American* 143, no. 5 (December 1930): 430. In most cases the pilot's ability to remain oriented in space went haywire, and the instruments reflected the actual but unrecognized behavior of the aircraft.

55. Kuhn, *Structure*, 169.

56. Myers, "The Medical Contribution," 92.

57. Ocker and Myers spent hours verifying their results in actual flight conditions; Ocker flew, and Myers, under a hood, signaled to Ocker what he thought the aircraft was doing by pulling on strings. Their conclusion was the same: the vestibular illusions "do take place in the air and, in addition, are much intensified." They also contacted the National Advisory Committee for Aeronautics (NACA), which conducted its own test flights and confirmed the findings of Ocker and Myers. See Myers, "The Medical Contribution," 94.

58. Jones, "Blind Flying," 74.

59. Hegenberger, "The Importance of the Development of Aerial Navigation," 2.

60. Eugene G. Reinartz, "Aviation and the Doctor," *Texas Reports on Biology and Medicine* 3, no. 3 (Fall 1945): 390. Brigadier General Reinartz served as the commandant of the School of Aviation Medicine from 1941 to 1946.

61. *Solving the Problem of Fog Flying: A Record of the Activities of the Fund's Full Flight Laboratory to Date* (New York: Daniel Guggenheim Fund for the Promotion of Aeronautics, 1929). Special manuscript series 539, box 3, Clark Special Collections Branch, USAFA Library.

62. The fund had ten trustees. Albert A. Michelson received the 1907 Nobel Prize in Physics for his study of the speed of light; Robert A. Millikan received the 1923 Nobel Prize in Physics for determining the charge of an electron. *Equipment Used in Experiments to Solve the Problem of Fog Flying: A Record of the Instruments and Experience of the Fund's Full Flight Laboratory* (New York: Daniel Guggenheim Fund for the Promotion of Aeronautics, 1930), 4. Special manuscript series 539, box 3, Clark Special Collections Branch, USAFA Library.

63. For further analysis of the Guggenheim Fund, see Richard P. Hallion, *Legacy of Flight: The Guggenheim Contribution to American Aviation* (Seattle: University of Washington, Press, 1977).

64. W. B. Courtney, "They're Coming through the Ceiling," *Colliers* 92, no. 12 (16 September 1933): 12.

65. *Solving the Problem of Fog Flying*, 15.

66. *Equipment Used in Experiments to Solve the Problem of Fog Flying*, 9.

67. Ibid., 11.

68. David Mindell, *Digital Apollo: Human and Machine in Spaceflight* (Cambridge, MA: MIT Press, 2008), 110. The Guggenheim Fund investigated several types of altimeters. The radio altimeter, a precursor to the modern radar altimeter, beamed radio waves toward the ground and assessed aircraft altitude according to how much time elapsed before the signal returned. An acoustic altimeter operated on the same principle. These instruments, however, remained too large and unreliable for installation on aircraft in the late 1920s. See "Solving the Problem," 28–31. The first test of the Kollsman altimeter occurred in 1928 when its inventor, Paul Kollsman, held it in his lap in the front seat of the Guggenheim Fund airplane while Jimmy Doolittle piloted from the rear seat. See Lieutenant General James H. Doolittle, interview by Lieutenant Colonel Robert M. Burch, Major Ronald R. Fogleman, and Captain James P. Tate, 26 September 1971. Oral History Collection, file 68, USAFA Library.

69. Doolittle earned his Ph.D. in 1925 from the Massachusetts Institute of Technology. For a useful analysis of Doolittle's pioneering work in aviation development during the 1920s, see Dik Alan Daso, *Doolittle, Aerospace Visionary* (Washington, DC: Brassey's, 2003).

70. Memorandum from Lieutenant James H. Doolittle to the Chief of the Air Service, Washington, DC, 19 August 1924. RG 18, box 2780, National Archives, College Park, MD.

71. Ibid.

72. A 1929 copy of the fund's official history contains numerous "handwritten comments by A. F. Hegenberger." See *Solving the Problem of Fog Flying.*

73. The Army focused some resources on instrument advances that occurred prior to the Guggenheim Fund. In 1924 the Engineering Division at McCook Field complained to the National Research Council that the development of new instruments was "hampered by the lack of a science of instrument engineering. Due to the lack of essential information, instruments must be developed largely by the 'cut and try' method." To improve the scientific basis of instrument development, the Engineering Division paid $10,900 to the Bureau of Standards for basic research in instrument mechanics. Letter from Assistant Chief, Engineering Division, to National Research Council, New York City, 2 February 1924. See "Development and Adaptation of Aircraft Instruments for Military Use," 12. File G-2, box 848, USAF Museum.

74. Albert F. Hegenberger and C. D. Barbulesco, "The Air Corps System of Instrument Landing (Type A-1)," Materiel Division, Wright Field, Dayton, OH, 1933. File G-2, box 849, USAF Museum.

75. Memorandum from F. Trubee Davison, Assistant Secretary of War, to Major General James E. Fechet, Chief of the Air Corps, 15 July 1930. Special manuscript series 539, box 3, Clark Special Collections Branch, USAFA Library.

76. Hegenberger and Barbulesco, "The Air Corps System."

77. "Instruments and Laboratory Equipment: General—Air Corps Instrument Landing System," *U.S. Army Air Corps Technical Order*, no. 05-1-5, (5 November 1935): 3, 11, 16. Special manuscript series 539, box 3, Clark Special Collections Branch, USAFA Library.

78. Albert F. Hegenberger and Fred H. Coleman, "Fog or Blind Landing System," *Air Corps Technical Report*, 4 March 1933. Special manuscript series 539, box 3, Clark Special Collections Branch, USAFA Library.

79. William Mitchell to Albert Hegenberger, 8 June 1932, call number 168.7279, iris number 01105916, AFHRA. After Mitchell resigned in 1926, he continued to proselytize about airpower. Notably, Jimmy Doolittle led a formation of B-25 "Mitchell" bombers over Tokyo in 1942. In 1946 Mitchell received a Special Congressional Medal of Honor as a posthumous recognition of his pioneering ideas regarding the air weapon.

80. William Mitchell to Albert Hegenberger, 27 February 1934, call number 168.7279, iris number 01105916, AFHRA.

81. Albert Hegenberger to William Mitchell, 14 March 1934, call number 168.7279, iris number 01105916, AFHRA.

82. Seventh Annual Report of the Chief, Materiel Division, US Army Air Corps, fiscal year 1933. Box R1-357, History Office, Air Force Materiel Command (AFMC), Wright-Patterson Air Force Base, OH.

83. Eighth Annual Report of the Chief, Materiel Division, US Army Air Corps, fiscal year 1934. Box R1-357, History Office, AFMC.

84. Hegenberger and Barbulesco, "The Air Corps System." By the end of 1934, the Bureau of Air Commerce placed an order for blind-landing equipment sets from the Air Corps. The goal was to equip twelve airports with this technology. See "Department of Commerce to Survey Sites at Twelve Airports for Installation of Blind Approach Equipment," *Air Commerce Bulletin* 6, no. 12 (15 June 1935): 277.

85. W. Irving Glover, "The Air Mail," *Annals of the American Academy of Political and Social Science* 131, Aviation (May 1927): 43. The US Post Office refined its airmail system throughout the 1920s, and its efforts to gain greater efficiencies in air transport highlighted the requirement for pilots to fly and navigate at night and through inclement weather. Erik M. Conway identifies the early airmail service as a significant catalyst in the development of blind-flight technology. For further insights on the relation between the US Post Office and US Army aviation, see Erik M. Conway, *Blind Landings: Low-Visibility Operations in American Aviation, 1918–1958* (Baltimore: Johns Hopkins University Press, 2006).

86. Bernard C. Nalty, ed., *Winged Shield, Winged Sword: A History of the United States Air Force*, vol. 1 (Washington, DC: Air Force History and Museums Program, 1997), 125.

87. James P. Tate, *The Army and Its Air Corps: Army Policy toward Aviation, 1919–1941* (Maxwell Air Force Base, AL: Air University Press, 1998), 131.

88. Rebecca Hancock Cameron, *Training to Fly: Military Flight Training, 1907–1945* (Washington, DC: Air Force History and Museums Program, 1999), 268. For more on the airmail fiasco's impact on instrument training in commercial and military aviation, see Conway, *Blind Landings*, 31–33.

89. "Instrument Flying," *U.S. Air Corps U-Stencils* 24, no. U-1144 (1935–36): 2. RG 18, entry 211, National Archives, College Park, MD.

90. Eighth Annual Report of the Chief, Materiel Division, U.S. Army Air Corps, fiscal year 1934. Box R1-357, History Office, AFMC.

91. *Air Corps Circular* 50-1 (22 March 1935). Special manuscript series 539, box 3, Clark Special Collections Branch, USAFA Library.

92. "Instrument Flying." Also, "Instrument Trainer Instruction Guide," AAF Technical Order No. 30-100C-1, 1 August 1943. Special manuscript series 603, box 2, Clark Special Collections Branch, USAFA Library. For more on the 1-2-3 method, see Conway, *Blind Landings*, 26–28.

93. Ninth Annual Report of the Chief, Materiel Division, US Army Air Corps, fiscal year 1935. Box R1-357, History Office, AFMC.

94. "Basic Instrument Flying," *U.S. Army Air Forces Technical Order*, no. 30-100A-1 (1 June 1943): 1–2. File G-2, Box 851, USAF Museum.

95. Ibid., 4.

96. "Instruments and Laboratory Equipment: General—Air Corps Instrument Landing System," *U.S. Army Air Corps Technical Order*, no. 05-1-5 (5 November 1935): 22. Special manuscript series 539, box 3, Clark Special Collections Branch, USAFA Library.

97. "Instrument Flying."

98. Similarly, in his analysis of the Apollo program, David Mindell relates the development of Apollo's flight computer to the engineering work of the 1930s that "helped change the nature of flight from 'seat of the pants' intuition to numerical, instrument-based tasks." See Mindell, *Digital Apollo*, 7.

99. "Basic Instrument Flying," 11.

100. Erik M. Conway, "The Politics of Blind Landing," *Technology and Culture* 42, no. 1 (January 2001): 82.

101. "Case History of AAF Navigational Aids, Part I: XT-1, Mark I GCA, AN/MPN-1 Series," Historical Office, Intelligence, T-2, Air Materiel Command, Wright Field, OH, September 1947. RG 18, entry 22, National Archives, College Park, MD.

102. Ibid., 2.

103. See Conway, "The Politics of Blind Landing," for a useful description of the basics of GCA.

104. Memorandum from Brigadier General B. W. Chidlaw, Assistant Chief of Staff, to the Director of Military Requirements, Director of Technical Services, Materiel Command, 23 January 1943. Also, see memorandum from Colonel Stuart F. Wright to the Director of Communications, HQ AAF, 29 January 1943. RG 18, entry 22, National Archives, College Park, MD.

105. "Case History of AAF Navigational Aids," 13.

106. Ibid., 2.

107. Conway, *Blind Landings*, 185.

108. George V. Holloman, "Automatically Controlled Blind Landings," *S.A.E. Journal* 42, no. 6 (June 1938): 13.

109. Speech by Major General Oscar Westover to the Western Aviation Planning Conference, 4 September 1937, Sacramento, CA. *U.S. Air Corps U-Stencils* 25, no. U-1203 (1937–38): 7.

CHAPTER FOUR: The Changing Role of the Human Component

1. In his 2002 book, *Between Human and Machine*, David Mindell focused on early twentieth-century "control systems," or technology that was "suffused with what would later be called

cybernetic ideas." His study emphasized naval fire control systems, the Sperry Gyroscope Company's contribution to the development of control systems, and early efforts in computing at the Bell Telephone Laboratories and MIT. His analysis informs this book in terms of how human components were integrated into larger technological systems, and his insights apply to the concept of aircraft as cybernetic systems. See David Mindell, *Between Human and Machine: Feedback, Control, and Computing before Cybernetics* (Baltimore: Johns Hopkins University Press, 2002). For more on the usefulness of cybernetics in understanding the modern era, see Ronald R. Kline, *The Cybernetics Moment: Or Why We Call Our Age the Information Age* (Baltimore: Johns Hopkins University Press, 2015).

2. Norbert Wiener, *Cybernetics, or Control and Communication in the Animal and the Machine* (New York: John Wiley and Sons, 1948), 19. The term *cyborg* was coined by Manfred Clynes and Nathan Kline in 1960. They fused the terms *cybernetic* and *organism* to describe "cyborgs," or modified humans that could operate in extreme environments with a minimum of external technological support. Similarly, Donna Haraway describes a cyborg as a "hybrid creature, composed of organism and machine," and the *Oxford English Dictionary* defines cyborg as "an integrated man–machine system." These definitions suggest that the pilot–aircraft cybernetic system could be termed a "cyborg" since it was a single entity interweaving organic and inorganic components. Nevertheless, this study prefers the term *cybernetic system*. The term *cyborg* invites confusion, since readers may associate it with contemporary notions of androids, Terminators, and other science fiction characters. See Manfred Clynes and Nathan Kline, "Cyborgs and Space," *Astronautics* 14 (1960): 26–27, 74–76; and Donna J. Haraway, *Simians, Cyborgs, and Women: The Reinvention of Nature* (New York: Routledge, 1991), 1.

3. Wiener, *Cybernetics*, 115.

4. Ibid., 36. In a speech to the National Advisory Committee for Aeronautics in 1949, physicist John D. Trimmer elaborated on the definition of cybernetics. It was "the science of observation, communication, and control in the animal and the machine." Observation involved the sensing of information by instruments or organic sense organs; communication was the transfer of information by mechanical devices or nerves; and control referred to the transformation of information into action by servomechanisms or muscles. See John D. Trimmer, "Instrumentation and Cybernetics," *Scientific Monthly* 69, no. 5 (November 1949): 328–31.

5. Peter Galison, "The Ontology of the Enemy: Norbert Wiener and the Cybernetic Vision," *Critical Inquiry* 21, no. 1 (Autumn 1994): 228–66. Galison provides a useful analysis of Norbert Wiener's activity in World War Two and its relevance to the future conceptualization of cybernetics.

6. Mindell, *Between Human and Machine*, 278. See also Wiener, *Cybernetics*, 11–13.

7. Galison, "The Ontology of the Enemy," 236. Galison concludes, "The core lesson that Wiener drew from his antiaircraft work was that the conceptualization of the pilot and gunner as servomechanisms within a single system was essential and irreducible" (238, 240).

8. Mindell, *Between Human and Machine*, 92–93, 98.

9. Galison, "The Ontology of the Enemy," 250–51.

10. German aeromedical researchers during World War Two assessed the nature of human control of aircraft and described the process of sensing and responding to information as a "control cycle." Similar to cybernetics, control cycles served as models of human interaction with mechanical components and informed aviation development. See Ulrich K. Henschke and Hans A. Mauch, "How Man Controls," *German Aviation Medicine, World War II*, vol. 1 (Washington, DC: US Government Printing Office, 1950), 83–91.

11. Wiener, *Cybernetics*, 19.

12. Mindell, *Between Human and Machine*, 74. Mindell offers a detailed account of the development and implementation of the Sperry Gyro Pilot, to include the "human steersman as a weak link in the system" and Sperry's characterization of his Gyro Pilot as a far more reliable wheelsman. See 71–76.

13. "Summary of Aviation Accident Reports from Training Centers, American Expeditionary Forces, Dec. 1917–May 1, 1918," 19. RG 18, library file B63 (15), box 115, National Archives, College Park, MD.

14. Mindell, *Between Human and Machine*, 138.

15. The Air Corps Technical School's Department of Mechanics described the autopilot as "a combination of gyroscopes and pneumatic and hydraulic actuated mechanisms which . . . simulate the operation of the brain, the nerves, and the muscles, respectively, of a human. The type A-2 automatic pilot is similar to the human body, but it acts faster on the controls, through its 'Brain,' 'Nerve,' and 'Muscular' system, than does the human body." "Aircraft Instruments," Air Corps Technical School Department of Mechanics, Personnel Training Department, 1 September 1939, San Antonio Air Depot, TX. File G-2, box 850, USAF Museum, Wright-Patterson Air Force Base, OH.

16. "Pilot Training Manual for the B-17 Flying Fortress," HQ Army Air Forces, Office of Flying Safety, 183. Special manuscript series 603, box 3, Clark Special Collections Branch, USAFA Library, Colorado Springs, CO.

17. James H. Doolittle, Lieutenant General, USAF (ret.), interview by E. M. Emme and W. D. Putnam, 21 April 1969, Washington, DC. File 68A, USAFA Library Oral History Collection.

18. E. P. Wheaton, "Human Factors in Engineering Design," Engineering Paper No. 903A, Douglas Aircraft Company, 20 January 1960, 5. Air University Library, Maxwell Air Force Base, AL.

19. Donald W. Hastings, Major, US Army Air Forces, et al., "Psychiatric Experiences of the Eighth Air Force, First Year of Combat (July 4, 1942–July 4 1943)." Call number 168.7082-642, iris number 1023197, AFHRA, Maxwell Air Force Base, AL.

20. Churchill Eisenhart, "Cybernetics: A New Discipline," *Science* 109, no. 2834 (22 April 1949): 397.

21. Major General Benjamin D. Foulois, USAF (Ret), "Early Flying Experiences in Army Airplane No. 1 (1909-1910-1911)," unpublished article, 1960. Manuscript series 17, box 6, Clark Special Collections Branch, USAFA Library. Foulois went on to become the chief of the Army Air Corps from 1931 to 1935.

22. Walter F. Grether, "The Genesis of Human Engineering," *Aviation, Space, and Environmental Medicine* 57, no. 10 (October 1986): A38. Dr. Grether was the chief of the Aero Medical Laboratory's Psychology Branch from 1949 to 1956. The term *human factors* is often substituted for human factors engineering. Grether also noted that various other terms were used in addition to human factors engineering. These included *engineering psychology, human engineering, biomechanics*, and *ergonomics*. Frank Hawkins, an airline pilot and the human factors consultant to KLM Royal Dutch Airlines, defined human factors as an applied technology that served to "optimize the relationship between people and their activities by the systematic application of the human sciences, integrated within the framework of systems engineering." See Frank H. Hawkins, *Human Factors in Flight* (Brookfield, VT: Gower Publishing, 1987), 18. E. P. Wheaton, the vice-president of engineering, missiles, and space systems at Douglas Aircraft Company, described human factors engineering as "the analysis of human behavioral capabilities and tolerances, and the integration of this information into the development of the most effective man–machine system." See E. P. Wheaton, "Human Factors in Engineering

Design," Engineering Paper No. 903A, Douglas Aircraft Company, 20 January 1960, 2. Air University Library, Maxwell Air Force Base, AL.

23. A 1958 article in the *Journal of Aviation Medicine* declared that "the need for rigorous man–machine considerations was not widely recognized until World War Two when the major impetus for application and research in this area was provided." The record of aviation history before the war suggests, however, that pilots, engineers, physicians, and military figures were well aware of human factors-related issues and made significant efforts to improve the man–machine relationship. Sherwin J. Klein and Charles F. Gell, "Aviation Human Engineering Is a Scientific Specialty," *Journal of Aviation Medicine* 29, no. 2 (March 1958): 213. For a further discussion of the development of human factors engineering and how flight surgeons facilitated technological changes to accommodate humans, see Maura Phillips Mackowski, "Human Factors: Aerospace Medicine and the Origins of Manned Space Flight in the United States" (Ph.D. diss., Arizona State University, 2002).

24. Walter G. Vincenti, *What Engineers Know and How They Know It* (Baltimore: Johns Hopkins University Press, 1990), 53. An aeronautical engineer and historian, Walter G. Vincenti analyzed how the interwar engineering community "learned to identify pilots' needs and translate them into criteria specifiable in terms appropriate to the hardware." 51. David Mindell's analysis of human–machine interactions in interwar aviation makes a similar point and describes how research engineers in the 1920s "measured the forces a pilot exerts on the cockpit controls and how they translate into an airplane's motion." See David Mindell, *Digital Apollo: Human and Machine in Spaceflight* (Cambridge, MA: MIT Press, 2008), 27.

25. *Handbook of Instructions for Airplane Designers*, 2nd ed., vol. 1 (Dayton, OH: US Army Air Service Engineering Division, 1921), 28. History Office, Air Force Materiel Command (AFMC), Wright-Patterson Air Force Base, OH.

26. Memorandum from Thurman H. Bane, Major, Chief of Engineering Division at McCook Field, to the Chief of the Air Service, Washington, DC, 22 November 1921. RG 18, library file 452.1, box 2782, National Archives, College Park, MD.

27. "Engineering," Army Air Service Advanced Flying School, Kelly Field, Texas, 1925. This was a text for the curriculum in advanced flying training. Special manuscript series 299, box 1, Clark Special Collections Branch, USAFA Library.

28. Edward C. Schneider, "The Human Machine in Aviation," *Air Service Information Circular* 5, no. 462, (1 May 1924): 1. Schneider was a lieutenant colonel in the Sanitary Officers Reserve Corps.

29. Wiener, *Cybernetics*, 13.

30. "Comparative Tests of Types L-1, Mark XI and D-7 Bombsights," *Air Corps Technical Report*, serial no. 3589, 4 February 1932. Call number 216.2101-12, iris number 00144235, AFHRA.

31. "Tests of Modified Air Corps Bombsight, Type D-4," Air Corps Technical Report, 7 November 1931. Call number 216.2101-12, iris number 00144231, AFHRA. And, L-1 Bombsight," Air Corps Technical Report, 10 May 1932. Call number 216.2101-12, iris number 00144234, AFHRA.

32. "Comparative Tests of Types L-1, Mark XI and D-7 Bombsights."

33. For a comprehensive analysis on the relationship between World War Two bomber aircraft crewmembers and automatic control technology, see Raymond P. O'Mara, "The Sociotechnical Construction of Precision Bombing: A Study of Shared Control and Cognition by Humans, Machines, and Doctrine during World War II" (Ph.D. diss., Massachusetts Institute of Technology, 2011).

34. Wiener, *Cybernetics*, 36.

35. *Handbook of Instructions for Airplane Designers*, 7th ed., vol. 1 (Wright Field, OH: US Army Air Corps, 1934), 317. History Office, AFMC.

36. Ibid., 318.

37. "The N-3A Gunsight, a Physical and Mental Hazard to the Fighter Pilot," memorandum from Captain Robert D. Mooney, Flight Surgeon, to the Chief Surgeon, Eleventh Air Force, 17 November 1942. RG 18, entry 22, box 51, National Archives, College Park, MD.

38. "Installation of K-14 Gyro Gunsight in P-51D Aircraft," memorandum from Colonel T. A. Sims, Chief of Administration, to Procurement Division and Engineering Division, Wright Field, 30 April 1945. RG 18, entry 22, box 51, National Archives, College Park, MD.

39. Grether, "The Genesis of Human Engineering," A38. Dr. Grether was the chief of the Psychology Branch from 1949 to 1956.

40. Rebecca J. Green, Herschel C. Self, and Tanya S. Ellefritt, eds., *50 Years of Human Engineering: History and Cumulative Bibliography of the Fitts Human Engineering Division* (Wright-Patterson Air Force Base, OH: Armstrong Laboratory, Air Force Materiel Command, 1995), 1–14, 1–15.

41. In a 1946 essay, Fitts argued: "If airplanes are to be operated with safety and effectiveness, they should be designed in relation to the psychological capacities and limitations of flyers as well as to their physiological requirements." Paul M. Fitts, "Psychological Requirements in Aviation Equipment Design," *Journal of Aviation Medicine* 17, no. 3 (June 1946): 270. In 1951 Fitts edited a frequently cited report on the air traffic control system that further explored the strengths and weaknesses of human and machine components. See Paul M. Fitts, ed., "Human Engineering for an Effective Air-Navigation and Traffic-Control System," a report to the Air Navigation Board and the National Research Council Committee on Aviation Psychology, March 1951.

42. J. L. Birley, "Report on the Medical Aspects of High Flying," *Reports of the Air Medical Investigation Committee*, 23 March 1918. RG 18, library file B63, box 115, National Archives, College Park, MD.

43. Harry G. Armstrong, "The Influence of Aviation Medicine on Aircraft Design and Operation," *Journal of the Aeronautical Sciences* 5, no. 5 (March 1938): 197.

44. Malcolm C. Grow and Harry G. Armstrong, *Fit to Fly: A Medical Handbook for Fliers* (New York: D. Appleton-Century, 1941), 13. Armstrong argued that "the best airplane in the world will be impotent against a much inferior machine if the pilot of the perfect airplane does not have the necessary 'stuff.' It requires only a shade of variation in human performance in the few short seconds of aerial encounter to make all the difference between victory or death."

45. Jacqueline Cochrane, "Final Report on Women Pilot Program," 1, Manuscript series 31, Clark Special Collections Branch, USAFA Library. The date is missing, but this report was submitted after the WASPs were deactivated on 20 December 1944. Cochrane, the director of Women Pilots, submitted this report to General Hap Arnold, chief of the Army Air Forces. She also noted that WASPs did "ferrying, target towing, tracking and searchlight missions, simulated strafing, smoke laying and other chemical missions, radio control flying, basic instrument instruction, engineering test flying, administrative and utility flying" (2).

46. General Henry "Hap" Arnold, speech at the Women's Air Force Service Pilots ceremony in Sweetwater, Texas, 7 December 1944. Manuscript series 31, Clark Special Collections Branch, USAFA Library.

47. Cochrane, "Final Report on Women Pilot Program," 39–40. Cochrane reported that incidents of flying fatigue were significantly lower in women. No explanation was given, however. Possible external factors relevant to this observation were that combat or the threat

of future combat compounded fatigue, and male aircrew flew longer missions as a function of their combat training. Fatigue, moreover, defied precise definition. See Harry G. Armstrong, "The Influence of Aviation Medicine on Aircraft Design and Operation," *Journal of the Aeronautical Sciences* 5, no. 5 (March 1938): 194.

48. R. E. Whitehead, M.D., "Notes from the Department of Commerce: Women Pilots," *Journal of Aviation Medicine* 5 (March–December 1934): 48.

49. Ibid., 83.

50. Raymond S. Holtz, M.D., "Should Women Fly during the Menstrual Period?" *Journal of Aviation Medicine* 12, no. 3 (September 1941): 302.

51. Cochrane, "Final Report on Women Pilot Program," 38.

52. David Meister, *The History of Human Factors and Ergonomics* (Mahwah, NJ: Lawrence Erlbaum Associates, 1999), 6–7. In his 2006 manuscript on the Apollo Program, David Mindell noted that the human is "a subject that engineering has never fully mastered." Mindell, *Digital Apollo*, 20.

53. "Summary of the Airplane Accidents of the Regular Army and the Reserve Corps," 1 July 1928–December 31, 1928. Call number 200.3912-2, iris number 00147790, AFHRA.

54. *Accident Bulletin for Medical Investigations, Continental U.S. Army Air Forces*, Office of Flying Safety, Medical Safety Division, Headquarters AAF, January 1945–January 1946, July 1945 monthly report, 1. Call number 259.3-7, iris number 00168715, AFHRA.

55. *Accident Bulletin for Medical Investigations*, June 1945 monthly report, 1.

56. James J. Hudson, *Hostile Skies: A Combat History of the American Air Service in World War I* (Syracuse, NY: Syracuse University Press, 1996), 299.

57. "Airplane Crashes, 1918 and 1919, in the United States Army Air Service," *Air Service Information Circular* 4, no. 340 (1 May 1922): 1.

58. "Human Factors in Accidents," report by the Department of Psychology, School of Aviation Medicine, Air University, US Air Force, October 1950. Air University Library.

59. Roland R. Birnn, "A War Diary," *Air Power Historian* 4, no. 2 (April 1957): 41.

60. Harold R. Tittman, personal letter, Special manuscript series 84, Clark Special Collections Branch, USAFA Library.

61. Peter Galison contended that accident investigators were inclined to find the causal source of commercial accidents in some form of pilot error. Unless mechanical failure was the obvious culprit, it was easy to shift blame onto the pilot since he or she possessed agency and was prone to inefficiency or error. Furthermore, finding a specific cause served the institutional interest of ascribing responsibility to an identifiable component of the system. See Galison, "An Accident of History," in *Atmospheric Flight in the Twentieth Century*, ed. Peter Galison and Alex Roland (London: Kluwer Academic Publishers, 2000), 3–44.

62. "Statistical Data on Aircraft Accidents and Forced Landings, FY 1932," Inspection Division, Office of the Chief of the Air Corps, September 1932. Call number 200.3912-2, iris number 00141796, AFHRA.

63. "Aircraft Accidents: Method of Analysis," Report No. 357, Committee on Aircraft Accidents, National Advisory Committee for Aeronautics, 17 January 1930. Call number 200.3912-2, iris number 00141756, AFHRA.

64. "Airplane Crashes, 1918 and 1919, in the United States Army Air Service," 1.

65. The National Advisory Committee for Aeronautics, Special Committee on the Nomenclature, Subdivision, and Classification of Aircraft Accidents, "The Classification of Aircraft Accidents: Nomenclature and Subdivision," 11 June 1928, 1. Call number 200.3912-2, iris number 00141796, AFHRA.

66. Ibid., 4. Similarly, Major General Mason Patrick, the chief of the Army Air Service from 1921 to 1927, wrote in 1928: "A very recent study of accidents" illustrated that "the

greatest number are due to the human equation, some error committed by the pilot." Mason M. Patrick, *The United States in the Air* (Garden City, NY: Doubleday, Doran, 1928), 145.

67. "The Classification of Aircraft Accidents," 6–7.

68. "Summary of the Airplane Accidents of the Regular Army and the Reserve Corps," 1 July 1928–December 31, 1928.

69. Ibid.

70. Ibid.

71. "Aircraft Accidents: Method of Analysis," Report No. 308, Special Committee on the Nomenclature, Subdivision, and Classification of Aircraft Accidents, National Advisory Committee for Aeronautics, 15 August 1928, 13. Call number 200.3912-2, iris number 00141755, AFHRA.

72. Ibid., 14.

73. Ibid.

74. H. Graeme Anderson, "The Medical Aspects of Aeroplane Accidents," *Aeronautics* 1, no. 43 (13 February 1918): 4.

75. Peter Galison argued that "the drive to regain control over the situation, to present recommendations for the future, to lodge moral and legal responsibility all urge the narrative towards a condensed causal account." He also noted that accidents occur in the context of a complex technological system shaped by external forces. The effort to find a sole cause may overlook other necessary causes in the chain of events leading to the mishap. Galison, "An Accident of History," 34, 21.

76. The NACA committee consisted of eight members in 1930: two from the Army Air Corps, two from the Navy, one from the Department of Commerce, and three from NACA. The definitions of pilot error and materiel failure remained unchanged from the 1928 model. "Aircraft Accidents: Method of Analysis," Report No. 357, 15.

77. "Aircraft Accidents: Method of Analysis," Report No. 308, 7.

78. "Technical Report of Aircraft Accident Classification Committee," 8 February 1934. Call number 200.3912-1, iris number 00139939, AFHRA.

79. "Aircraft Accidents: Method of Analysis," Report No. 357, 7.

80. Paul M. Fitts and Richard E. Jones, "Analysis of Factors Contributing to 460 Pilot Error Experiences in Operating Aircraft Controls," Report No. TSEAA-694-12, Psychology Branch, Aero Medical Laboratory, 1 July 1947, 25. Document number ADB814586, Defense Information Systems Agency, Defense Technical Information Center, Ft. Belvoir, VA. As noted earlier, Dr. Fitts was the first chief of the Psychology Branch. He began this report in December 1945 and relied on written reports as well as aircrew interviews.

81. The documentation in this report does not specify which category of pilot error was identified. According to the preceding definitions, it may have been labeled pilot error due to "carelessness or negligence."

82. Fitts and Jones, "Analysis of Factors," 25.

83. Grether, "The Genesis of Human Engineering," A40.

84. Fitts and Jones, "Analysis of Factors," 4.

85. Ibid., 22–23. Raising the flaps reduced the lift provided by the wings.

86. *Accident Bulletin for Medical Investigations*, October 1945 monthly report, 4.

87. For analyses of World War Two airpower strategy and its development in the interwar years, see Tami Davis Biddle, *Rhetoric and Reality in Air Warfare* (Princeton, NJ: Princeton University Press, 2002); David E. Johnson, *Fast Tanks and Heavy Bombers* (Ithaca, NY: Cornell University Press, 1998); Geoffrey Perret, *Winged Victory: The Army Air Forces in World War II* (New York: Random House, 1993); Michael S. Sherry, *The Rise of American Airpower* (New Haven,

CT: Yale University Press, 1987); and Mark K. Wells, *Courage and Air Warfare: The Allied Aircrew Experience in the Second World War* (London: Frank Cass, 1995).

88. The Eighth Air Force, based in England, was the Army Air Forces organization tasked with conducting the massive strategic bombing operations in Western Europe.

89. Letter from Major General Ira C. Eaker to Major General George E. Stratemeyer, 30 January 1943. Manuscript series 33, box 30, Clark Special Collections Branch, USAFA Library.

90. Memorandum from Major General H. A. Craig, Assistant Chief of Air Staff for Operations, Commitments, and Requirements, to the Assistant Chief of Air Staff for Plans, 21 August 1944. RG 18, entry 294, box 380, National Archives, College Park, MD.

91. Sheila Jasanoff notes that others have criticized the "transformation of the human from godlike inventor to cog in the machine as one of modern technology's worst unintended consequences." Sheila Jasanoff, "Technology as a Site and Object of Politics," in *The Oxford Handbook of Contextual Political Analysis*, ed. Robert E. Goodin and Charles Tilly (Oxford University Press, 2006), 746.

92. Galison, "The Ontology of the Enemy," 236; Mindell, *Between Human and Machine,* 92–93, 98.

93. Henry H. Arnold and Ira C. Eaker, *This Flying Game* (New York: Funk and Wagnalls, 1936), 86.

94. Ibid., 91.

95. "Survival at Altitude of Heavy Bomber Crews," Second Air Force Altitude Indoctrination Unit, July 1944. File G-3, box 859, USAF Museum.

96. "Instrument Flying Training," *Technical Manual 1-445* (Washington, DC: War Department, 1942), 2.

97. *Accident Bulletin for Medical Investigations*, May 1945 monthly report, 3.

98. "Standardization of Army Primary Flying Training," *Supplement to Training Manual TM 1-210* (Maxwell Field, AL: Army Air Forces Southeast Training Center, 1943), 1. File G-2, box 848, USAF Museum. This anticipated one historian's observation that, "to the extent that technology orders or designs the physical and psychological parameters of human existence, it does so through sometimes forcible processes of standardization that demarcate normal social identities and behaviors from those regarded as deviant or abnormal." See Sheila Jasanoff, "Technology as a Site and Object of Politics," in *The Oxford Handbook of Contextual Political Analysis*, ed. Robert E. Goodin and Charles Tilly (Oxford: Oxford University Press, 2006), 747.

99. Wiener, *Cybernetics*, 13.

100. Fitts and Jones, "Analysis of Factors," 11.

101. For a summary of this accident, see Phillip S. Meilinger, "When the Fortress Went Down," *Air Force Magazine* 87, no. 10 (October 2004): 78–82.

102. Donald L. Putt, Lieutenant General, USAF (ret.), interview by Murray Green, 13 August 1974, Atherton, CA. Manuscript series 33, box 9, Clark Special Collections Branch, USAFA Library.

103. "Boeing Test Chief Dies of Injuries," *New York Times*, 20 November 1935, p. 6.

104. "Pilot Training Manual for the B-17 Flying Fortress," Headquarters Army Air Forces, Office of Flying Safety (n.d.). Special manuscript series 603, box 3, Clark Special Collections Branch, USAFA Library. The foreword to this manual contained a message from the chief of the Army Air Forces, General Henry Arnold, that emphasized the concept of standardization: "The techniques and procedures described in this book are standard and mandatory. . . . [This is] a complete exposition of what your pilot duties are, how each duty will be performed, and why it must be performed in the manner prescribed."

105. "Standardization of Army Primary Flying Training," 6.

106. "The B-29: Airplane Commander Training Manual for the Superfortress," *Army Air Forces Manual 50-9* (Washington, DC: HQ AAF, Office of Flying Safety, 1 February 1945), 24–33. Special manuscript series 1279, box 1, Clark Special Collections Branch, USAFA Library.

107. Frederick Winslow Taylor, *The Principles of Scientific Management* (New York: W. W. Norton, 1967). 39–40, 64.

108. Ibid., 36.

109. Ibid., 7; Wiener, *Cybernetics*, 38.

110. Anson Rabinbach, *The Human Motor: Energy, Fatigue, and the Origins of Modernity* (Berkeley: University of California Press, 1992): 239.

111. Speech by Brigadier General Henry H. Arnold to the Western Aviation Planning Conference, 4 September 1937, Sacramento, CA. *U.S. Air Corps U-Stencils* 25, no. U-1203 (1937–38): 17.

112. The terms *technological trajectory* and *pattern of change* used here are informed by historian Donald MacKenzie and economist Giovanni Dosi. For MacKenzie, technological trajectories "exist in the sense of persistent patterns of technological change." Persistence was governed by self-perpetuating, institutional interests. See Donald MacKenzie, *Inventing Accuracy: A Historical Sociology of Nuclear Missile Guidance* (Cambridge, MA: MIT Press, 1990), 168. Dosi employed the term technological trajectory to describe the continuities of change in a technological system or paradigm. For Dosi, a technological paradigm echoed Kuhn's concept of a scientific paradigm. It was "an outlook, a set of procedures, a definition of the 'relevant' problems and of the specific knowledge related to their solution" within which technological trajectories could develop. Giovanni Dosi, "Technological Paradigms and Technological Trajectories," *Research Policy* 11 (1982): 148.

113. Army aviation gained some experience before World War One. The Signal Corps' First Aero Squadron, commanded by Captain Benjamin Foulois, provided airborne reconnaissance and communication support during the 1916 expedition against Pancho Villa.

114. "Sperry Automatic Pilot," *Aero Digest* 15, no. 6 (December 1929): 166. Ships used gyroscopes to aid stability in rough seas.

115. Most aircraft of the early twentieth century lacked longitudinal or lateral stability and required constant vigilance on the part of their pilots. On the basis of experiments with gliders and before achieving powered flight, Wilbur Wright suggested that pilots would have to possess the personality type that is willing to ride a "fractious horse." Wilbur Wright, "Some Aeronautical Experiments," speech to the Western Society of Engineers, 18 September 1901, http:// www.wright-house.com/wright-brothers/Aeronautical.html.

116. For a thorough account of Sperry's efforts to develop gyroscope-based stabilization in ships and aircraft, see Thomas P. Hughes, *Elmer Sperry: Inventor and Engineer* (Baltimore: Johns Hopkins University Press, 1971), 103–53, 173–200, and Mindell, *Between Human and Machine*, 71–76.

117. "Automatic Pilot Installation in Twin-Motored Hydro-Aeroplane," *Aerial Age Weekly* 4, no. 16 (1 January 1917): 409.

118. *Solving the Problem of Fog Flying: A Record of the Activities of the Fund's Full Flight Laboratory to Date* (New York: Daniel Guggenheim Fund for the Promotion of Aeronautics, 1929), 16. Special manuscript series 539, box 3, Clark Special Collections Branch, USAFA Library.

119. The Sperry autopilot could sense rotation of one-half of one degree around the vertical, lateral, or longitudinal axes of the airplane. "Sperry Automatic Pilot," *Aero Digest* 15, no. 6 (December 1929): 166.

120. Captain Albert Hegenberger to Colonel William Mitchell (USAAC, ret.), 14 March 1934, call number 168.7279, iris number 01105916, AFHRA. The 14 March letter was Hegenberger's

reply to Mitchell's 27 February request: "If you have any data available on instrument flying, automatic pilots, landing fogs, resonance or capacity altimeters, defrosting arrangements or directional radio equipment, I would very much like to see it."

121. Wiener, *Cybernetics*, 13.

122. Fifth Annual Report of the Chief, Materiel Division, US Army Air Corps, fiscal year 1931, 102. Box R1-357, History Office, Air Force Materiel Command (AFMC), Wright-Patterson Air Force Base, OH. The Sperry A-1 flew the C-7A for fourteen hours during a test in 1931.

123. Ibid., 93.

124. Ninth Annual Report of the Chief, Materiel Division, US Army Air Corps, fiscal year 1935, 37. Box R1-357, History Office, AFMC.

125. Circular Letter No. 37-21, Office of the Chief of the Air Corps, 10 November 1937. File G-2, box 849, USAF Museum.

126. "Part Played by the Air Corps in Connection with Instrument Flying," *U.S. Air Corps U-Stencils* 25, no. U-1189 (1937–38): 234. RG 18, entry 211, box 13, National Archives, College Park, MD.

127. George V. Holloman, "Automatically Controlled Blind Landings," *S.A.E. Journal* 42, no. 6 (June 1938): 13.

128. Circular Letter No. 37-21.

129. Memorandum from Major General Delos C. Emmons, Commanding General HQ Air Force, to the Chief of the Air Corps, 14 July 1939. RG 18, entry 241-A, box 3, National Archives, College Park, MD.

130. Report on the 27 July 1939 board at Langley Field. RG 18, entry 241-A, box 3, National Archives, College Park, MD.

131. Memorandum from Lieutenant General Delos C. Emmons, Commander, HQ Air Force, to Major General Henry "Hap" Arnold, Chief of the Air Corps, 8 April 1941. RG 18, entry 241-A, box 3, National Archives, College Park, MD (emphasis original).

132. Churchill considered the raid "a brilliant exploit, the effectiveness of which the photographs have revealed." Eaker responded, "We will repeat these efforts many times, and on an ever-increasing scale." "Bombs Away," *Time* 41, no. 13 (29 March 1943).

133. Report from the Office of the Assistant Chief of Air Staff, Intelligence, Washington, DC, 13 April 1943, 1. Call number 142.052, iris number 00115849, AFHRA.

134. Memorandum from Headquarters, Second Air Division, to Commanding General, Eighth Air Force, 12 February 1945. Call number 526.804B, iris number 00229756, AFHRA.

135. Report from the Office of the Assistant Chief of Air Staff, 4.

136. Ibid., 5.

137. For more perspective on the relationship between the pilot, bombardier, and automatic control mechanisms, see O'Mara, "The Socio-technical Construction of Precision Bombing," 326–29.

138. Report from the Office of the Assistant Chief of Air Staff, 5.

139. The standardization of aircrew performance carried its own price. While it enabled greater accuracy during bombing runs, it also presented a predictable target to enemy defenses.

140. "Automatic Flight Control Equipment," Operations Analysis Section, IX Bomber Command, Ninth US Air Force, 14 September 1943. Call number 131.504f, iris 0111936, AFHRA.

141. William R. Stark, "The Gadget Did the Trick," *Air Force* 27, no. 3 (March 1944): 22. Call number 245.606, iris 00156507, AFHRA.

142. The tension between human decision making and the checklist's machine-like scripting of human action illustrates one aspect of assimilating the pilot into a complex technological

system that optimizes safety and effectiveness. For an instructive analysis of how this played out between astronauts, engineers, and spacecraft in the Apollo Program, see Mindell, *Digital Apollo*, 14.

143. "The B-29: Airplane Commander Training Manual for the Superfortress," 7.

144. For more on the role of how the B-17 pilot "managed the system at both the individual machine and larger system levels," see O'Mara, "The Socio-technical Construction of Precision Bombing," 339.

CHAPTER FIVE: Flight without Flyers

1. Sidney Shalett, "Arnold Reveals Secret Weapons, Bomber Surpassing All Others," *New York Times*, 18 August 1945.

2. Memorandum from Major Henry H. Arnold, Materiel Division, to the Chief, Plans Division, Office of the Chief of the Air Corps, 18 August 1930, Exhibit A, 1. RG 18, entry 22, box 55, National Archives, College Park, MD.

3. Delmar Fahrney and Robert Strobell, "America's First Pilotless Aircraft," *Aero Digest* 66, no. 1 (July 1954): 28; H. R. Everett, *Unmanned Systems of World Wars I and II* (Cambridge, MA: MIT Press, 2015), 248–53. For another summary of the development of the aerial torpedo, see Thomas P. Hughes, *American Genesis: A Century of Invention and Technological Enthusiasm, 1870–1970* (Chicago: University of Chicago Press, 1989), 126–34.

4. "Summary of Power Driven Weapons," Special Weapons Branch, Equipment Laboratory, Wright Field, OH, 20 November 1943. Call number 201.46, iris 00141374, AFHRA, Maxwell Air Force Base, AL. The initial Navy version was an N-9 seaplane that carried a pilot as a safety observer. It could also carry 1,000 pounds of explosive up to fifty miles. Sperry demonstrated its automatic flying capabilities in 1917 with a pilot on board. On 6 March 1918, this device was launched without a pilot and flew straight ahead for 1,000 yards until the engine stalled after a predetermined number of revolutions. The plane then spiraled into the water. Fahrney and Strobell, "America's First Pilotless Aircraft," 29.

5. Arnold memorandum to the Chief, Plans Division, Exhibit A, p. 1.

6. Fahrney and Strobell, "America's First Pilotless Aircraft," 30.

7. "The Kettering Bug," *History of USAF Drone/RPV* (Wright-Patterson Air Force Base, OH, 1976). File C2, box 655, USAF Museum, Wright-Patterson Air Force Base, OH.

8. Fahrney and Strobell, "America's First Pilotless Aircraft," 30.

9. Arnold memorandum to the Chief, Plans Division, Exhibit A, p. 2.

10. Fahrney and Strobell, "America's First Pilotless Aircraft," 30.

11. Memorandum Report, "Aerial Torpedo," Major Henry H. Arnold, Materiel Division, to Chief of the Air Corps, 18 August 1930, 1. RG 18, entry 22, box 55, National Archives, College Park, MD.

12. Ibid.

13. Arnold memorandum to the Chief, Plans Division, Exhibit A, p. 4.

14. Letter from Major Thurman H. Bane, Commanding Officer, McCook Field, to Major Henry H. Arnold, Ninth Corps Area, San Francisco, 1 September 1921. Manuscript Series 33, box 26, Clark Special Collections Branch, USAFA Library, Colorado Springs, CO.

15. Arnold memorandum to the Chief, Plans Division, Exhibit A, p. 8.

16. "Summary of Power Driven Weapons," 2.

17. Arnold memorandum to the Chief, Plans Division, Exhibit A, p. 11.

18. Memorandum from Major General Mason M. Patrick, Chief of the Air Service, to Engineering Division, McCook Field, 4 August 1923. This memo was titled, "Aerial Torpedoes for Bombing Battleships." RG 18, decimal file 452.1, box 2792, National Archives, College

Park, MD. The highly publicized sinking of the German battleship *Ostfriesland* by Air Service bombers occurred in 1921, and additional tests against warships were conducted in 1923.

19. Memorandum from the Chief of the Engineering Division, McCook Field, to the Chief of the Air Service, 25 October 1924. RG 18, decimal file 452.1, box 2781, National Archives, College Park, MD. General Mitchell wrote in 1925 that "aerial torpedoes which are really airplanes kept on their course by gyroscopic instruments and wireless telegraphy [radio], with no pilots on board, can be directed for over a hundred miles in a sufficiently accurate way to hit great cities." William Mitchell, *Winged Defense* (New York: Dover Publications, 1988), 6, 165.

20. This appraisal was directed by the assistant secretary of War for Air, F. Trubee Davison, on 13 September 1927. Letter from the Chief of the Materiel Division to the Chief of the Air Corps, 10 October 1928. RG 18, entry 22, box 55, National Archives, College Park, MD.

21. Ibid.

22. Ibid.

23. Letter from Brigadier General William Gillmore, Chief, Materiel Division, to F. Trubee Davison, Secretary of War for Air, 12 April 1927. RG 18, entry 22, box 55, National Archives, College Park, MD.

24. For a seminal analysis of the relationship between organization, military doctrine, and technological development, see Irving B. Holley Jr., *Ideas and Weapons* (New Haven, CT: Yale University Press, 1953). Holley argued that the development of superior weapons was impeded by ineffective organizations that failed to "attach sufficient importance to the formulation of doctrine" (176).

25. Letter from the Chief, Materiel Division to the Chief of the Air Corps, 10 October 1928. RG 18, entry 22, box 55, National Archives, College Park, MD.

26. Arnold memorandum to the Chief, Plans Division, Exhibit C, p. 3.

27. Regarding limited assets, see "Radio Controlled Aircraft," Report of the Air Corps Board, Maxwell Field, AL, 25 October 1935, 4. Call number 167.5-4, iris 001211122, AFHRA. Regarding termination of the program, see "Detailed Progress Report on Aerial Torpedo Development," Engineering Section, Materiel Division, Wright Field, OH, 31 March 1934. RG 18, entry 22, box 55, National Archives, College Park, MD.

28. Memorandum from Major General Benjamin D. Foulois, Chief of the Air Corps, to the Adjutant General, 6 May 1935. Call number 167.5-4, iris 00121122, AFHRA.

29. The Coast Artillery Board and the commanding general, GHQ Air Force, also requested the chief of the Air Corps to initiate a study by the Air Corps Board. "Radio Controlled Aircraft," 1.

30. Ibid., 7. See also comments from the chief of the Materiel Division: Memorandum from Brigadier General A. W. Robins, Chief of Materiel Division, to the Chief of the Air Corps, 31 May 1935. Call number 167.5-4, iris 00121122, AFHRA.

31. The first page of the board's 25 October 1935 report included Westover's comment: "Approved in principle. However no commitment for procurement of commercial type airplanes to carry out experiments for radio controlled flight is authorized at this time." "Radio Controlled Aircraft," Report of the Air Corps Board, i.

32. Memorandum from Major General Westover to the Chief, Materiel Division, Wright Field, 8 August 1938. RG 18, entry 22, box 55, National Archives, College Park MD.

33. Memorandum from Major General Westover to the Chief, Materiel Division, Wright Field, 29 August 1938. RG 18, entry 22, box 55, National Archives, College Park MD.

34. Letter from Major Henry H. Arnold to Major Thurman H. Bane, 10 August 1921. Manuscript series 33, box 26, Clark Special Collections Branch, USAFA Library.

35. In 1939 Arnold wrote to Charles Kettering, inventor of the "Bug" during World War One, encouraging the revival of the aerial torpedo. Arnold noted that "our failure in the past to follow up in a logical manner the knowledge gained on the early experiments was due in part to the tendency to experiment for the sake of research, thereby losing sight of the original concept of the aerial torpedo which demanded that it be capable of being produced cheaply and quickly." Letter from Henry "Hap" Arnold to Charles F. Kettering, 3 November 1939, 2. RG 18, entry 22, box 55, National Archives, College Park, MD.

36. Memorandum from the Chief of the Air Corps to the Adjutant General, 29 September 1938. RG 18, entry 22, box 55, National Archives, College Park, MD.

37. Memorandum from the Adjutant General to the Chief of the Air Corps, 7 October 1938. RG 18, entry 22, box 55, National Archives, College Park, MD.

38. Memorandum from the Chief of the Air Corps to the Adjutant General, 8 October 1938. RG 18, entry 22, box 55, National Archives, College Park, MD.

39. General Arnold employed this strategy on other occasions as well. In the development of glide-bombs, or bombs that traveled more of a horizontal path toward targets after release from a bomber, Arnold envisioned precision weapons that could glide to pinpoint targets. The unguided versions possessed little military value, but they were "the first step in a development program that eventually should lead to precision bombing." Letter from Henry "Hap" Arnold to Lieutenant General Frank Andrews, Caribbean Defense Command, 8 September 1942. Call number 202.4-1, iris 00142198, AFHRA.

40. The Army approved $10,000 for a design competition and allotted another $90,000 to develop the winning design. Memorandum from the Adjutant General to the Chief of the Air Corps, 31 October 1938. RG 18, entry 22, box 55, National Archives, College Park, MD.

41. Memoranda from General Henry "Hap" Arnold, "Military Characteristics of Aerial Torpedoes," 15 and 23 January 1940. RG 18, entry 22, box 55, National Archives, College Park, MD.

42. Letter from Arnold to Kettering, 3 November 1939, 2.

43. Memorandum from the Commanding General, Materiel Command, to the Commanding General, Army Air Forces, and the Assistant Chief of the Air Staff, Materiel, Maintenance, and Distribution, "Controllable Bomb, Power-Driven, General Motors, Type A-1," 24 August 1943, 4. RG 18, entry 22, box 55, National Archives, College Park, MD.

44. Memorandum from the Bureau of Aeronautics, Navy Department, to Colonel Grandison Gardner, Army Air Corps, 18 April 1941. Call number 202.4-1, iris 00142199, AFHRA.

45. Memorandum, "Controllable Bomb, Power-Driven, General Motors, Type A-1," 4.

46. Ibid.

47. Rolling out of control due to turbulence resembled the sport of "V-1 tipping" described in chapter 4 where pilots destroyed V-1 cruise missiles by tipping them over and upsetting their gyros.

48. Letter from Colonel Grandison Gardner to Major General Henry "Hap" Arnold, 24 April 1942. RG 18, entry 22, box 55, National Archives, College Park, MD.

49. In March 1943, a GMA-1 using television guidance was crashed seventy-five feet from its target. Memorandum from Materiel Command to the Commanding General, Army Air Forces, 12.

50. Routing and Record Sheet, "Controllable Bomb Program," 18 December 1942. RG 18, entry 22, box 55, National Archives, College Park, MD. In this context, "Interstate" referred to weapons capable of traveling international distances.

51. Memorandum from Materiel Command to the Commanding General, Army Air Forces, 12. These particular weapons never came to fruition in World War Two. In 1942–43, the Army

Air Forces experimented with variants of large drones designated XBQ-1, XBQ-2, and XBQ-3. Expensive failures in early tests doomed these weapons in their initial experimental phase. For a summary of cruise missile development, see Kenneth P. Werrell, *The Evolution of the Cruise Missile* (Maxwell Air Force Base, AL: Air University Press, 1985).

52. Joseph P. Kennedy Jr. to John F. Kennedy, 10 August 1944, National Archives, http://www.media.nara.gov/media/images/31/27/31-2604a.gif.

53. "Joseph P. Kennedy, Jr.," http://www.jpkf.org/BIOG.HTML.

54. Memorandum from Colonel H. J. Knerr, Chief of Staff, GHQ Air Force, to the Chief of the Air Corps, 30 July 1935. Call number 167.5-4, iris 00121122, AFHRA.

55. Memorandum Report, "Controllable Bomb Projects," Lieutenant Colonel Randolph P. Williams, Experimental Engineering Section, Materiel Division, 13 December 1941, 3. RG 18, entry 22, box 55, National Archives, College Park, MD.

56. "Case History of Controlled Missiles—Aircraft, Part II—Castor," Historical Division, Air Materiel Command, 24 July 1945, 1. RG 18, entry 22, box 55, National Archives, College Park, MD.

57. Memorandum from Brigadier General B. W. Chidlaw, Chief, Materiel Division, to Major General O. P. Echols, Assistant Chief of Air Staff, Materiel, Maintenance, and Distribution, 1 March 1944, Appendix III, 3. Call number 201.46, iris 00141073, AFHRA.

58. Memorandum from Colonel R. C. Wilson, Chief, Developmental Engineering Branch, Materiel Division, to the Commanding General, Materiel Command, Wright Field, 8 June 1944. Call number 201.46, iris 00141374, AFHRA.

59. Wesley F. Craven and James L. Cate, eds., *The Army Air Forces in World War II*, vol. 3, *Europe: Argument to V-E Day, January 1944 to May 1945* (Washington, DC: Office of Air Force History, 1983), 531. See also Memorandum from Lieutenant Colonel Paul F. Davis, Chief, Special Projects Section, Office of Air Communications, to General Tom C. Rives, 21 October 1944. Titled "Use of War Weary Aircraft as Robots," this memo noted that "an urgent requirement was established by the USSTAF, approximately 1 July 44 for radio controlled heavy bombardment aircraft to be stripped and loaded to capacity with explosives or incendiary for use against targets in the Pas de Calais area." RG 18, entry 22, box 38, National Archives, College Park, MD. General Henry "Hap" Arnold noted with approval on 1 July 1944 General Spaatz's desire to send Castor B-17s against German launch sites. See Jack Olsen, *Aphrodite: Desperate Mission* (New York: G. P. Putnam's Sons, 1970), 46.

60. Although the TDR-1 "torpedo drone" and its 2,000-pound payload demonstrated promising results in combat against Japanese positions in September 1944, the Navy canceled the program in October 1944. Everett, *Unmanned Systems of World Wars I and II*, 321–41. For documentary footage, see "Service Test in Field of TDR1—WWII, Torpedo Drone 30770," www.youtube.com/watch?v=8RQcUtzAe98.

61. Memorandum from Vice Admiral Aubrey W. Fitch, Chief of Naval Operations, to General Henry "Hap" Arnold, Commanding General, Army Air Forces, 20 September 1944. Call number 202.4-1, iris 00142197, AFHRA. Admiral Fitch noted the Navy "has carried on extensive development in the powered aircraft assault drone field" and produced nearly 200 small drones "for various test purposes." In praising the logic of the Army Air Forces' development of heavy-bomber assault drones, he noted the Navy's cancellation of its drone operations and transfer of control equipment and some smaller drones to the Army Air Forces for further development and employment of this technology.

62. Olsen, *Aphrodite: Desperate Mission*, 109–11.

63. Fain H. Pool, interview transcript from "The Voices of WWII," Bristol Productions, 1998, http://wwiihistoryclass.com/transcripts/Pool_F_012.pdf. Pool piloted the Aphrodite B-17 on 11 September 1944. See also Olsen, *Aphrodite: Desperate Mission*, 109–11.

64. Communiqué from Lieutenant General James Doolittle to General Carl Spaatz, 15 December 1944. Call number 519.1622, iris 00214864, AFHRA.

65. Ibid.

66. Memorandum Report, "Castor Project," Engineering Division, Air Technical Service Command, 13 January 1945. RG 18, entry 22, box 38, National Archives, College Park, MD. These buildings were not the primary target at the Heligoland complex, but the gratifying amount of local destruction enabled the mission to count as a success.

67. Memorandum from Colonel S. R. Brentnall, Assistant Chief, Materiel Division, to Director, Army Air Forces Air Technical Service Command, Wright Field, 17 October 1944. RG 18, entry 22, box 55, National Archives, College Park, MD. See also communiqué from Lieutenant General James Doolittle to General Carl Spaatz, 9 November 1944. Call number 519.1622, iris 00214864, AFHRA.

68. "Guided Missiles Developed by the Army Air Forces, 1940–1945, Part II," 9. This was a secret document produced by air force historians at the end of the war. RG 18, entry 22, box 55, National Archives, College Park, MD.

69. Memorandum from Brigadier General F. O. Carroll, Chief, Engineering Division, to General Henry "Hap" Arnold, Commanding General, Army Air Forces, 19 January 1945. RG 18, entry 22, box 38, National Archives, College Park, MD.

70. Memorandum to Headquarters, Eighth Air Force, 8 December 1944. Call number 519.1622, iris 00214864, AFHRA.

71. Ibid.

72. Routing and Record Sheet from Air Communications Officer, Special Projects Section, Materiel Division, 27 September 1944. RG 18, entry 22, box 38, National Archives, College Park, MD. See also "Case Histories of Controlled Missiles—Aircraft, Part II—Castor," 2–3.

73. Memorandum from Brigadier General Grandison Gardner, Commander, AAF Proving Ground Command, Eglin Field, FL, to General Arnold, Commander, Army Air Forces, 30 September 1944. Call number 202.4-1, iris 00142197, AFHRA.

74. Communiqué from General Henry "Hap" Arnold to General Carl Spaatz, 11 November 1944. Call number 519.1622, iris 00214864, AFHRA.

75. Communiqué from General Carl Spaatz to General Henry "Hap" Arnold, 21 November 1944. Call number 519.1622, iris 00214864, AFHRA.

76. Memorandum from Brigadier General Donald Wilson, Assistant Chief of Staff Air Staff, to General Henry "Hap" Arnold, Commander, Army Air Forces, 22 December 1944. Call number 202.4-1, iris 00142197, AFHRA.

77. The Air Technical Service Command at Wright Field conducted numerous tests with radar-guided B-17s and B-24s to determine the best way to crash or "dump" them into their targets: "Various schemes for 'dumping' the airplane have been proposed, such as blowing off the wings or tail surfaces, stalling the airplane and subsequently spinning it into the target, and steep-angle dives into the target." Tests indicated that the best procedure was to cut the engines and let the aircraft glide into its target. Memorandum Report, "Trajectories of B-17 and B-24 Missile Aircraft," Air Technical Service Command, Wright Field, 20 January 1945. AD number ADB813927, Defense Technical Information Center (DTIC), Ft. Belvoir, VA.

78. Memorandum from Headquarters Ninth Air Force to General Carl Spaatz, 22 November 1944. Call number 519.1622, iris 00214864, AFHRA.

79. "Long Range, High Speed, Ground to Ground, Pilotless Aircraft," 1 June 1945. AD number ADB807746, DTIC.

80. Memorandum from the Headquarters Army Air Forces, Requirements Division, to the Assistant Chief of Air Staff Operations, Commitments, and Requirements, 7 February 1945, 2. RG 18, entry 22, box 38, National Archives, College Park, MD.

81. The Army Air Forces considered the feasibility of employing Willie Baby against Japan. Favorable weather enabled B-29 Mother aircraft to find targets for B-24 Babies, and the Babies could also be used to create a diversion for Japanese air defenses. Nevertheless, the Army Air Forces preferred to conduct more tests, and the plan was never executed. "Army Air Forces Board Preliminary Report on Willie (Baby) Aircraft controlled from B-29s on Long Range Missions," 15 December 1944, 3–4. Call number 202.4-1, iris 00142197, AFHRA.

82. "Project 4396A 471.6," Army Air Forces Board, Orlando, FL, 27 February 1945. See "Abstract Report and Supplements," Guided Missiles Committee, 26 April 1946. Call number 178.264-2, iris 00131921, AFHRA.

83. Winston S. Churchill, *The Second World War: Triumph and Tragedy* (Boston: Houghton Mifflin, 1953), 39.

84. Ibid.

85. For a summary of V-1 employment, see Werrell, *The Evolution of the Cruise Missile*, 41–62.

86. Memorandum from Brigadier General F. O. Carroll, Chief, Engineering Division, Materiel Command, to the Deputy Chief of Staff, Materiel Command, 12 July 1944. Call number 201.46, iris 00141375, AFHRA.

87. Memorandum from Major General H. A. Craig, Assistant Chief of Air Staff, Operations, Commitments, and Requirements, to President, Army Air Forces Board, 24 August 1944. Call number 201.46, iris 00141375, AFHRA.

88. Ibid.

89. Communiqué from USSTAF to Supreme Headquarters Allied European Forces, 18 December 1944. Another communiqué between the same organizations on 26 December 1944 noted that an expansion to 500 JB-2 launches per day would cost the equivalent of 1,250,000 105-millimeter high explosive rounds or 22,9000,500 pounds of general purpose bombs. "Guided Missiles Developed by the Army Air Forces, 1940–1945, Part II." RG 18, entry 22, box 55, National Archives, College Park, MD.

90. Memorandum from Brigadier General E. M. Powers, Deputy Assistant Chief of Staff, Materiel and Services, to Director, Air Technical Service Command, 14 January 1945. Call number 201.46, iris 00141376, AFHRA. For data on the number of V-1s launched, see Werrell, *The Evolution of the Cruise Missile*, 60.

91. Memorandum from Brigadier General E. M. Powers, 14 January 1945.

92. Communiqué from USSTAF to Supreme Headquarters Allied European Forces, 18 December 1944.

93. Letter from Lieutenant General Barney M. Giles, Deputy Commander, Army Air Forces and Chief of Air Staff, to Dr. Vannevar Bush, Office of Scientific Research and Development, 13 February 1945. Call number 201.46, iris 00141376, AFHRA. For a similar account, see Werrell, *The Evolution of the Cruise Missile*, 79.

94. "Special Weapons for Operations against Japan," Joint Committee on New Weapons and Equipment, Guided Missiles Committee, Subcommittee Number 3, 1 May 1945, 13. Call number 201.46, iris 00141376, AFHRA. For an analysis of how principles of strategic bombing compared to actual practice and how the strategy of the Army Air Forces changed from precision to indiscriminate bombing during World War Two, see Tami Davis Biddle, *Rhetoric and Reality in Air Warfare* (Princeton, NJ: Princeton University Press, 2002).

95. "Guided Missiles Developed by the Army Air Forces, 1940–1945, Part II," 43.

96. Memorandum from Major General H. A. Craig, 24 August 1944.

97. Memorandum Report, "Guided Missiles," Lieutenant General Barney M. Giles, Chief of the Air Staff, 6 February 1945. RG 18, entry 22, box 56, National Archives, College Park, MD. General Giles noted the essential need "to gain air superiority by attacks against aircraft

in the air and on the ground and against those installations which the enemy requires for the application of air power." He also observed that, "The JB-2 and any weapon with similar capabilities can conceivably be employed in accomplishing any or all of the [air force's] missions."

98. Mitchell, *Winged Defense*, x.

99. Theodore von Kármán, *The Wind and Beyond* (Boston: Little, Brown, 1967), 298.

100. Memorandum from Robert A. Lovett, Assistant Secretary of War for Air, to General Henry "Hap" Arnold, 2 August 1944. Call number 201.46, iris 00141073, AFHRA. Secretary Lovett claimed that this proposed weapon would be "a self-propelled missile . . . capable of being controlled in speed, azimuth, elevation, and range to within 100 yards of any selected target."

101. Memorandum from Dr. Edward L. Bowles, Expert Consultant to the Secretary of War, to Colonel S. F. Giffin, Executive Requirements Division, Assistant Chief of Air Staff, 7 August 1944. Call number 201.46, iris 00141073, AFHRA.

102. Memorandum from Colonel S. F. Giffin, Executive Requirements Division, Assistant Chief of Air Staff, Operations, Commitments and Requirements, to Dr. Edward L. Bowles, Expert Consultant to the Secretary of War, 11 August 1944. Call number 201.46, iris 00141073, AFHRA.

103. Memorandum from Colonel William F. McKee, Assistant Chief of Air Staff, Operations, Commitments and Requirements, to the Deputy Chief of Air Staff, 18 January 1945. Call number 201.46, iris 00141073, AFHRA. Similarly, the Materiel Command contended that guided missiles "should be the responsibility of the Army Air Forces when: the missile is essentially an aircraft; though not essentially an aircraft is nevertheless capable of sustained flight controlled by aerodynamic or aerostatic means; the missile is controlled from an aircraft; [the missile] is controlled by some device normally assigned to and employed by the Army Air Forces; the missile is susceptible to use as an alternative or additional bombing weapon." Memorandum from Colonel J. F. Phillips, Materiel Command, to the Chief of the Air Staff, 5 September 1944. Call number 201.46, iris 00141073, AFHRA.

104. Memorandum from Lieutenant General Joseph T. McNarney, Deputy Chief of Staff, US Army, to the Commanding General, Army Air Forces, 2 October 1944. Call number 201.46, iris 00141073, AFHRA. Kenneth Werrell provided an alternative account to this intraservice dispute and noted, "The McNarney memo divided responsibility on an evolutionary basis: winged missiles looked and performed like aircraft and therefore went to the AAF, wingless missiles looked and performed like artillery and, hence, went to the ASF [Army Service Forces]." Werrell, *The Evolution of the Cruise Missile*, 80.

105. John Catchpole, *Project Mercury: NASA's First Manned Space Programme* (Chichester, UK: Praxis Publishing, 2003), 32–34.

106. "Navy and Ordnance Guided Missile Projects, Part III," 12. RG 18, entry 22, box 55, National Archives, College Park, MD.

107. "The Development of Guided Missiles," Historical Division, Air Materiel Command, June 1946, 27. RG 18, entry 22, box 55, National Archives, College Park, MD.

108. Memorandum from Major General Alden Crawford, Air Force Director of Research and Engineering, to Deputy Chief of Staff, US Army, 21 February 1946. Call number 201.46, iris 00141073, AFHRA.

109. Memorandum from Colonel S. F. Giffin, 11 August 1944.

110. Record and Routing Sheet from Major General H. R. Oldfield, Special Assistant for Antiaircraft, to Assistant Chief of the Air Staff, Operations, Commitments and Requirements, 27 November 1944. Call number 201.46, iris 00141073, AFHRA. See also Robert F. Futrell,

Basic Thinking in the United States Air Force, 1907–1960, vol. 1 (Maxwell Air Force Base, AL: Air University Press, 1989), 202–3.

111. For much of the Second World War the Materiel Command's Special Weapons Branch helped the Army Air Forces develop pilotless weapons. The branch's engineers worked on remote-control equipment for the Castor Project, for example, as well as other devices such as winged bombs that glided several miles to their targets after aerial release and homing devices based on radar or infrared radiation. Major General O. P. Echols, Assistant Chief of Air Staff, to Chief of the Air Staff [no date given, but written before 15 August 1944]. Call number 201.46, iris 00141073, AFHRA. For a summary of air force missile development, see Jacob Neufeld, "Ace in the Hole," in *Technology and the Air Force, A Retrospective Assessment*, ed. Jacob Neufeld, George M. Watson Jr., and David Chenoweth (Washington, DC: Air Force History and Museums Program, 1997), 111–15.

112. Memorandum from Colonel W. S. Roth, Chief, Aircraft Projects, Service Engineering Section, Engineering Division, to Commanding General, Army Air Forces, 16 February 1945. RG 18, entry 22, box 55, National Archives, College Park, MD.

113. Major General St. Clair Street to General Henry "Hap" Arnold, "Navy and Ordnance Guided Missile Projects, Part II." RG 18, entry 22, box 55, National Archives, College Park, MD.

114. Shalett, "Arnold Reveals Secret Weapons," 4.

115. Ibid.

116. This reflects historian Lynn White's notable claim that a new technical ability "merely opens a door; it does not compel one to enter. The acceptance or rejection of an invention, or the extent to which its implications are realized if it is accepted, depends quite as much upon the condition of a society, and upon the imagination of its leaders, as upon the nature of the technological item itself." Lynn White Jr. *Medieval Technology and Social Change* (London: Oxford University Press, 1968), 28.

117. Memorandum from Brigadier General L. C. Craigie, Chief, Engineering Division, Air Technical Service Command, to Major General Leslie R. Groves, Corps of Engineers, 6 December 1945. RG 18, entry 22, box 56, National Archives, College Park, MD. This memo referred to the use of winged cruise missiles, not ballistic missiles. For further analysis on the postwar development of cruise missiles, see Werrell, *The Evolution of the Cruise Missile*, 79–128. For insights into the development and guidance of nuclear-tipped ballistic missiles, see Edmund Beard, *Developing the ICBM: A Study in Bureaucratic Politics* (New York: Columbia University Press, 1976), and Donald MacKenzie, *Inventing Accuracy: A Historical Sociology of Nuclear Missile Guidance* (Cambridge, MA: MIT Press, 1990).

118. The Army Air Forces also sought to advance its mastery of cutting-edge technology by obtaining German scientists at the end of the war. This effort was called Project Paperclip and recruited German experts in rocketry, nuclear physics, chemistry, and aviation medicine. For more on Paperclip, see Maura Phillips Mackowski, "Human Factors: Aerospace Medicine and the Origins of Manned Space Flight in the United States" (Ph.D. diss., Arizona State University, 2002), 188–239.

CHAPTER SIX: The Modern Pilot, Redefined

1. Air Corps test pilot Captain George Holloman noted in 1937 that delegating cockpit workload to automatic devices likened the pilot to an efficient manager of an increasingly complex system. See George V. Holloman, "Automatically Controlled Blind Landings," *S.A.E. Journal* 42, no. 6 (June 1938): 13. Indeed, the phenomenon of pilots becoming more like managers of systems and less like actuators of controls reflects the increasing complexity of

aviation. Historian David Mindell cites the 1959 remarks of Elwood Quesada, a retired Air Force general then installed as the first administrator of the Federal Aviation Administration, in an address to the Society of Experimental Test Pilots. Despite favoring a dominant role for pilots, he presciently noted, "The day of the throttle jockey is past. He is becoming a true professional, a manager of complex weapons systems." David Mindell, *Digital Apollo: Human and Machine in Spaceflight* (Cambridge, MA: MIT Press, 2008), 40.

2. Rob Schapiro, "A Dangerous Problem with Airline Pilot Training," *American Thinker*, 28 May 2011, http://www.americanthinker.com/blog/2011/05/a_dangerous_problem_with_airli.html.

3. Arnold Reiner, "Pilots on Autopilot," *New York Times*, 16 December 2009.

4. Dave Majumdar, "AF: UAV Pilots Will Eventually Outnumber Others," *Air Force Times*, 19 September 2011. This article reflects the observation of General Edward Rice, the commander of the US Air Force's Air Education and Training Command.

5. "Post's Sleep on Flight Put at Only 20 Hours," *New York Times*, 23 July 1933. David Mindell portrays the Sperry A-2 automatic pilot used by Post as a vital, interactive component: "The machine was not perfect; sometimes it failed and Post had to fly manually. The two worked together, trading control, playing on each other's strengths and weaknesses." See David Mindell, *Between Human and Machine: Feedback, Control, and Computing before Cybernetics* (Baltimore: Johns Hopkins University Press, 2002), 79–80.

6. "Post's Automatic Pilot," *New York Times*, 24 July 1933.

7. Anthony Leviero, "Robot-Piloted Plane Makes Safe Crossing of Atlantic: No Hand on Controls from Newfoundland to Oxfordshire—Take-Off, Flight and Landing Are Fully Automatic," *New York Times*, 22 September 1947. The *Times* also noted, "Air Force officers speculated on the possibility of loading robot planes, like the Skymaster, with bombs and sending them to distant targets. For peaceful uses, it was suggested they might be used as cargo carriers."

8. Report from aircraft crew and passengers, http://www.history.com/topics/charles-a-lindbergh/speeches/transatlantic-flight-by-autopilot.

9. Joan Lowy, "Automation in the Air Dulls Pilot Skill," *Associated Press*, 2 September 2011. See also William R. Nelson, James C. Byers, Lon N. Haney, Lee T. Ostrom, and Wendy J. Reece, Idaho National Engineering Laboratory, Lockheed Idaho Technologies, "Lessons Learned from the Introduction of Cockpit Automation in Advanced Technology Aircraft." This 1995 NASA-sponsored study concluded, "The introduction of sophisticated autopilot, flight control, and flight management systems on modern aircraft . . . has resulted in significant changes in the operation of these aircraft. The flight crews have shifted to more of a system manager role rather than that of in-the-loop pilots." http://www.osti.gov/scitech/servlets/purl/114648.

10. "Final Report on the Accident on 1st June 2009 to the Airbus A330-203 Registered F-GZCP Operated by Air France, Flight AF 447 Rio de Janeiro—Paris," Bureau d'Enquêtes et d'Analyses pour la sécurité de l'aviation civile, 27 July 2012, pp. 25, 27, 28. BEA is the French Civil Aviation Investigation Authority.

11. Ibid., 35.

12. Nicola Clark, "Report on '09 Air France Crash Cites Conflicting Data in Cockpit," *New York Times*, 5 July 2012.

13. BEA Report, 37 (emphasis added).

14. Ibid.

15. Ibid., 21.

16. Ibid., 36.

17. Ibid., 185.

18. Ibid., 176.

19. Ibid. The initial indications on the ECAM also informed the crew that the aircraft's autopilot was disconnected, the aircraft no longer possessed wind-shear detection, and the maximum permissible speed at the current altitude was 330 knots or 0.82 mach. This data made no reference to the failure of accurate airspeed sensors, the real cause of the sudden disconnect of the autopilot.

20. Ibid., 38.

21. William Langewiesche, *Fly by Wire: The Geese, the Glide, the Miracle on the Hudson* (New York: Farrar, Straus, and Giroux, 2009), 108.

22. The "alternate law" and "protections lost" indications reveal modern aviation's robust reliance on automation, a phenomenon with roots deep in the early history of flight.

23. BEA Report, 37.

24. Ibid., 180.

25. Ibid., 197.

26. Ibid., 181. "Indeed, the charged emotional factor combined with the workload prompted the PF [pilot flying] to trust the flight director, independently of any other parameter: he may have considered the flight director crossbars as means of maintaining the cruise level" (181). The BEA report concluded, "Even if it is not sure that the crew followed the orders from the flight director while the stall warning was active, the orders from the crossbars were in contradiction with the inputs to make in this situation and thus may have troubled the crew. Consequently, the BEA recommends that: EASA [European Aviation Safety Agency] require a review of the functional or display logic of the flight director so that it disappears or presents appropriate orders when the stall warning is triggered" (211).

27. Ibid., 199.

28. Ibid., 102.

29. Ibid., 31.

30. Ibid., 199.

31. Ibid., 204.

32. Ibid.

33. Ibid.

34. Ibid., 209.

35. Ibid., 211.

36. Clark, "Report on '09 Air France Crash Cites Conflicting Data in Cockpit."

37. http://www.faa.gov/aircraft/air_cert/design_approvals/csta/tech_discipline/flight_deck/.

38. Asaf Degani et al., "Modes in Automated Cockpits: Problems, Data Analysis, and a Modeling Framework," Proceedings of the 36th Israel Annual Conference on Aerospace Sciences, Haifa, Israel, 21–22 February 1996.

39. "Operational Use of Flight Path Management Systems," Final Report of the Performance-Based Operations Aviation Rulemaking Committee / Commercial Aviation Safety Team Flight Deck Automation Working Group, Federal Aviation Administration, 5 September 2013, pp. 2, 3, http://www.faa.gov/about/office_org/headquarters_offices/avs/offices/afs/afs400/parc/parc_reco/media/2013/130908_PARC_FltDAWG_Final_Report_Recommendations.pdf. For additional analysis, see David Mindell, *Our Robots, Ourselves* (New York: Viking Press, 2015), 73–77.

40. Chesley Sullenberger and Jeffrey Zaslow, *Highest Duty: My Search for What Really Matters* (New York: HarperCollins, 2009), 190. The FAA's chief scientific and technical advisor for flight deck human factors, Dr. Kathy Abbott, similarly notes, "Although the automatic systems have reduced or eliminated some types of pilot errors, the automatic systems have introduced other types of errors." http://www.faa.gov/aircraft/air_cert/design_approvals/csta/tech_discipline/flight_deck/.

41. Lowy, "Automation in the Air."

42. Earl L. Wiener, "Human Factors of Advanced Technology ('Glass Cockpit') Transport Aircraft," prepared for Ames Research Center, NASA, June 1989. In their three-year study of Boeing 757 crews, researchers noted, "many of the crews expressed the view that automation may have gone too far, that they felt they were often 'out of the loop,' probably meaning that they tended to lose situational awareness, and that they feared that automation led to complacency." http://www.ntrs.nasa.gov/search.jsp?R=19890016609.

43. Langewiesche, *Fly by Wire*, 132–33.

44. William Langewiesche, "Anatomy of a Miracle," *Vanity Fair*, no. 586 (June 2009).

45. National Transportation Safety Board Accident Report, "Loss of Thrust in Both Engines after Encountering a Flock of Birds and Subsequent Ditching on the Hudson River, US Airways Flight 1549, Airbus A320-214, N106US," NTSB/AAR-10/03, PB2010-910403, 4 May 2010, p. 88.

46. Langewiesche, *Fly by Wire*, 108.

47. Ibid., 99, 137.

48. Stephen P. Rosen, *Winning the Next War: Innovation and the Modern Military* (Ithaca, NY: Cornell University Press, 1991), 202.

49. Memorandum from General Arnold to Dr. Theodore von Kármán, 7 November 1944. Theodore von Kármán, *Toward New Horizons: A Report to General of the Army H. H. Arnold, Submitted on Behalf of the A.A.F. Scientific Advisory Group* (1945), v.

50. Norton Schwartz, General, USAF, Air Force Association Convention Speech, 15 September 2009, Washington, DC.

51. Edmund Beard, *Developing the ICBM: A Study in Bureaucratic Politics* (New York: Columbia University Press, 1976), 39. LeMay wrote in his capacity as the head of the US Army Air Forces Research and Development office.

52. Herman S. Wolk, *Fulcrum of Power: Essays on the United States Air Force and National Security* (Washington, DC: Air Force History and Museums Program, 2003), 115, http://www.dtic.mil/dtic/tr/fulltext/u2/a440080.pdf.

53. John Stillion, "Trends in Air-to-Air Combat: Implications for Future Air Superiority," Center for Strategic and Budgetary Assessments, Washington, DC, 2015. The CSBA describes itself as "an independent, nonpartisan policy research institute established to promote innovative thinking and debate about national security strategy and investment options."

54. Ibid., 24.

55. Ibid., i.

56. "Fly, fight, and win" is the stated mission and mantra of the US Air Force.

57. 90th Fighter Squadron Fact Sheet, http://www.jber.af.mil/library/factsheets/factsheet.asp?id=7713.

58. David Bedard, "Flagship: 90th Fighter Squadron Receives New F-22," Public Affairs article, Joint Base Elmendorf-Richardson, 8 March 2012, http://www.jber.af.mil/news/story.asp?id=123293241.

59. Ibid.

60. Stillion, "Trends in Air-to-Air Combat," 33. For a detailed analysis of how technological enhancements and automation have altered the skillset of fighter pilots, see Steven A. Fino, *Tiger Check: Automating the US Air Force Fighter Pilot in Air-to-Air Combat, 1950–1980* (Baltimore: Johns Hopkins University Press, 2017).

61. Douglas McIntyre and Michael Sauter, "The 10 Most Expensive Weapons in the World," *Wall Street Journal*, 9 January 2012, http://www.247wallst.com/special-report/2012/01/09/the-10-most-expensive-weapons-in-the-world/3/.

62. The estimate of $391 billion is from a Department of Defense Selected Acquisition Report released 19 March 2015 and reporting numbers for 31 December 2014, Release No. NR-090-15, www.defense.gov/Releases/Release.aspx?ReleaseID=17181. For Lockheed Martin's basic claims about standoff capabilities, see http://www.f35.com/about/capabilities.

63. Norbert Wiener, *Cybernetics, or Control and Communication in the Animal and the Machine* (New York: John Wiley and Sons, 1948), 36.

64. Christian Davenport, "Meet the Most Fascinating Part of the F-35: The $400,000 Helmet," *Washington Post*, 1 April 2015.

65. William B. Albery, "Multisensory Cueing for Enhancing Orientation Information during Flight," *Aviation, Space, and Environmental Medicine*, 78, no. 5, sec. II (May 2007): B186.

66. Kevin Gray, "The Last Fighter Pilot," *Popular Science*, January–February 2016, http://www.popsci.com/last-fighter-pilot.

67. Davenport, "Meet the Most Fascinating Part of the F-35."

68. Northrup Grumman video, "F-35 JSF Distributed Aperture System (DAS) Sensors Demonstrate Hostile Fire Detection Capability," 11 February 2013, http://www.youtube.com/watch?v=fHZO0T5mDYU&list=PLxYF2Xt6-JqGp-LHnQucGbtbQTBdsnFp2&index=2.

69. Northrop Grumman video, "F-35 Electro-Optical Distributed Aperture System," 13 May 2010, http://www.youtube.com/watch?v=e1NrFZddihQ.

70. "The F35B Pilot's New Helmet and DAS: A Huge Leap in Air-Ground Decision-Making Sharing," *Second Line of Defense* interview with Lieutenant Colonel Dehner, USMC, 28 May 2010, http://www.sldinfo.com/distributed-aperture-systems-das-and-the-f35b-pilots-new-helmet-re-shaping-tactical-capabilities/.

71. Marty Kauchak, "International Expansion," *Military Simulation and Training Magazine*, February 2015, p. 9. This observation comes from Mike Luntz, director of Lockheed Martin's F-35 Training System. For comparison, 72 percent of F-35 initial training sorties are flown in the simulator, whereas F-16 pilots fly only 40 percent of their sorties in a simulator.

72. Davenport, "Meet the Most Fascinating Part of the F-35."

73. Ibid.

74. "Second Addendum to United States Air Force Accident Investigation Board Report, F-22A, T/N 06-4125," 525th Fighter Squadron, 3rd Wing, Joint Base Elmendorf-Richardson, Alaska, 7 June 2013, http://usaf.aib.law.af.mil/ExecSum2011/F-22%202nd%20Addendum%2C%20JBER%2C%2016%20Nov%2010.pdf.

75. A Department of Defense Inspector General report on the F-22 crash criticized the Air Force's official report for inadequately examining the possible effects of hypoxia as a causal factor in this fatal mishap. The pilot's loss of awareness may therefore have been aggravated by a faulty human–machine interface. http://www.dodig.mil/pubs/documents/DODIG-2013-041.pdf.

76. "Second Addendum to United States Air Force Accident Investigation Board Report, F-22A, T/N 06-4125."

77. Jeremiah Gertler, "Air Force F-22 Fighter Program," Congressional Research Service, 11 July 2013, p. 11, http://www.govexec.com/pdfs/022511rb1.pdf.

78. Peter Merlin, "NASA-Pioneered Automatic Ground-Collision Avoidance System Operational," Public Affairs, NASA Armstrong Flight Research Center, 7 October 2014.

79. "General Aviation Controlled Flight into Terrain Awareness," *Advisory Circular*, no. 61-134, 1 April 2003, Federal Aviation Administration, p. 3.

80. Guy Norris, "Ground Collision Avoidance System 'Saves' First F-16 in Syria," *Aerospace Daily & Defense Report*, 5 February 2015, http://www.aviationweek.com/defense/ground-collision-avoidance-system-saves-first-f-16-syria.com. From 1992 to 2004, the USAF's F-16 fleet experienced 34 CFIT accidents with twenty-four fatalities. See Jet Fabara, "Keeping the

Warfighter Safe: 416 FLTS Continues to Enhance Life-Saving Auto Collision Avoidance Technology," 412th Test Wing Public Affairs, 20 November 2012, http://www.afmc.af.mil/news/story.asp?id=123327065.

81. David J. Niedober et al., "Influence of Cultural, Organizational and Automation Factors on Human-Automation Trust: A Case Study of Auto-GCAS Engineers and Developmental History," 16th International Conference on Human-Computer Interaction, Heraklion, Crete, 22 June 2014.

82. Fabara, "Keeping the Warfighter Safe."

83. Norris, "Ground Collision Avoidance System."

84. Niedober et al., "Influence of Cultural."

85. Guy Norris, "F-16 Flight Demonstrates Auto-GCAS Potential," *Aviation Week & Space Technology*, 2 August 2010, http://www.aviationweek.com/technology/f-16-flight-demonstrates-auto-gcas-potential. For details on NASA's development of Auto-GCAS for the F-16 as well as remote-controlled aircraft, see Peter Merlin, "NASA-Pioneered Automatic Ground-Collision Avoidance System Operational," Public Affairs, NASA Armstrong Flight Research Center, 7 October 2014, http://www.nasa.gov/centers/armstrong/Features/Auto-GCAS_Installed_in_USAF_F-16s.html.

86. Fabara, "Keeping the Warfighter Safe."

87. The researchers are from California State University Northridge's Systems Engineering Research Laboratory, NASA's Dryden Flight Research Center and Ames Research Center, and the Air Force's Research Laboratory and Flight Test Center. Niedober et al., "Influence of Cultural."

88. Ibid.

89. Ibid.

90. Thomas B. Sheridan and William L. Verplank, "Human and Computer Control of Undersea Teleoperators," Man–Machine Systems Laboratory, Department of Mechanical Engineering, MIT, 14 July 1978. Sheridan's work was informed by the influential aviation psychologist Paul M. Fitts and his 1951 report on the strengths and weaknesses of human and machine components in a complex technological system. See Paul M. Fitts, ed., "Human Engineering for an Effective Air-Navigation and Traffic-Control System," a report to the Air Navigation Board and the National Research Council Committee on Aviation Psychology, March 1951. See also Thomas B. Sheridan, *Telerobotics, Automation, and Human Supervisory Control* (Cambridge, MA: MIT Press, 1992). Caitlin Lee applies the Sheridan–Verplank taxonomy to remotely piloted aircraft in "Embracing Autonomy: The Key to Developing a New Generation of Remotely Piloted Aircraft for Operations in Contested Air Environments," *Air and Space Power Journal* 25, no. 4 (Winter 2011): 80.

91. Kolina Koltai et al., "Influence of Cultural, Organizational, and Automation Capability on Human Automation Trust: A Case Study of Auto-GCAS Experimental Test Pilots," International Conference on Human-Computer Interaction in Aerospace, Silicon Valley, CA, 30 July 2014.

92. Barrie Barber, "New Tech Tool at Wright-Patt Saving Lives," *Dayton Daily News*, 18 December 2015, http://www.mydaytondailynews.com/news/news/local-military/new-tech-tool-developed-at-wright-patt-saving-live/npnH4/.

93. Molly Lachance, Air Force Office of Scientific Research, "AFOSR Study Evaluates Trust in Autonomous Systems," 7 May 2015, http://www.wpafb.af.mil/news/story.asp?id=123447432.

94. James Alan Marshall, telephone interview by author, 25 June 2014. Colonel Marshall served as the US Air Force's Air Combat Command chief of safety and oversaw safety-related flight policy throughout the Air Force's bomber and fighter communities. He emphasized

the enduring applicability of this mantra and recalled how he drummed it into his students during his several years of service as a jet pilot instructor.

95. Mindell, *Digital Apollo*, 24, 27.

96. This also comports with David Mindell's observation that pilots, when faced with new opportunities for their professional development, are willing to "give up their attachment to total control" and develop skills beyond traditional stick-and-rudder mastery. Ibid., 24.

97. Mark A. Skoog and James L. Less, "Development and Flight Demonstration of a Variable Autonomy Ground Collision Avoidance System," paper presented at the American Institute of Aeronautics and Astronautics, Atlanta, Georgia, 16 June 2014, http://www.nasa.gov/sites/default/files/files/DR-0005-DRC-012-033_iGCAS-paper_2014-06-28.pdf.

98. Schwartz, Air Force Association Convention Speech, 15 September 2009.

CHAPTER SEVEN: New Horizons of Flight

1. In 1904 Ludwig Prandtl identified the "boundary layer," which explained how air flows over the surface of the wing. Prandtl was a professor of applied mechanics at Gottingen who later became a leading aerodynamicist. Theodore von Kármán, "Engineering Education in Our Age," *Quadrangle* 31, no. 2 (December 1960): 10. For more on the notion that science lagged behind technology in aviation, see John D. Anderson Jr., "The Evolution of Aerodynamics in the Twentieth-Century: Engineering or Science?" *Atmospheric Flight in the Twentieth Century*, ed. Peter Galison and Alex Roland (London: Kluwer Academic Publishers, 2000), 241–56.

2. Wiebe E. Bijker, Thomas P. Hughes, and Trevor Pinch, eds., *The Social Construction of Technological Systems: New Directions in the Sociology and History of Technology* (Cambridge, MA: MIT Press, 2012), 19.

3. Theodore von Kármán, *The Wind and Beyond* (Boston: Little, Brown, 1967), 271. Von Kármán quoted from a letter written to him by General Arnold.

4. Record of Long Distance Telephone Conversation between Major B. W. Chidlaw, Office Chief, Materiel Division, and Lieutenant Colonel F. O. Carroll, Wright Field, 7 March 1941. RG 18, entry 22, box 55, National Archives, College Park, MD.

5. Harold McGee, "The Triple Alliance: Millikan, Guggenheim, and von Kármán," *Engineering and Science* 44, no. 4 (April 1981): 26, http://www.calteches.library.caltech.edu/3287/.

6. James H. Doolittle, Lieutenant General, USAF (Ret.), interview by E. M. Emme and W. D. Putnam, 21 April 1969, Washington, DC. File 68A, USAFA Library Oral History Collection, Colorado Springs, CO.

7. Von Kármán, *The Wind and Beyond*, 225–26.

8. Ibid., 267–68.

9. Memorandum from Theodore von Kármán to General Henry Arnold, 9 January 1946. SMS 328, box 1, Clark Special Collections Branch, USAFA Library. Dr. von Kármán's recruits included some of his colleagues from the California Institute of Technology as well as aerodynamicist Dr. Hugh Dryden, Dr. Ivan Getting from the MIT Radiation Laboratory, and future Nobel laureate Dr. E. M. Purcell. See von Kármán, *The Wind and Beyond*, 269–70.

10. Von Kármán, *The Wind and Beyond*, 271. For more on the Scientific Advisory Board, see Thomas A. Sturm, *The USAF Scientific Advisory Board: Its First Twenty Years, 1944–1964* (Washington, DC: USAF Historical Division Liaison Office, 1 February 1967).

11. For more on the creation of the Scientific Advisory Board, see Dik Alan Daso, *Architects of American Air Supremacy: General Hap Arnold and Dr. Theodore von Kármán* (Maxwell Air Force Base, AL: Air University Press, 1997). Other sources that address General Arnold's efforts to

develop a science-based air force include Thomas M. Coffey, *Hap: The Story of the U.S. Air Force and the Man Who Built It, General Henry H. "Hap" Arnold* (New York: Viking, 1982); Dik Alan Daso, *Hap Arnold and the Evolution of American Air Power* (Washington, DC: Smithsonian Institution Press, 2000); and Flint O. DuPre, *Hap Arnold: Architect of American Air Power* (New York: Macmillan, 1972). While offering useful portrayals of General Arnold's dominant influence on American airpower, these biographies overlook the deep roots of science and scientific engineering that took hold in the decades before World War Two and provided the foundation for the air force's scientific outlook.

12. In a 15 December 1945 memorandum to General Arnold, Theodore von Kármán asserted, "The next ten years should be devoted to the realization of the potentialities of scientific progress, with the following principal goals: supersonic flight, pilotless aircraft, all-weather flying, perfected navigation and communication, remote-controlled and automatic fighter and bomber forces, and aerial transportation of entire armies." Theodore von Kármán, *Toward New Horizons: A Report to General of the Army H. H. Arnold, Submitted on Behalf of the A.A.F. Scientific Advisory Group* (Washington, DC, 1945), p. ix.

13. Ibid., 2.

14. Ibid., 5.

15. Ibid.

16. For a detailed exposition of the development of the aviation system, see Erik M. Conway, *Blind Landings: Low-Visibility Operations in American Aviation, 1918–1959* (Baltimore: Johns Hopkins University Press, 2006).

17. Some scholars express alarm at this notion. Historian Sheila Jasanoff, for example, argues that in many modern technological systems "the system's logic overrides its members' desire for self-expression. Many have deplored this transformation of the human from godlike inventor to cog in the machine as one of modern technology's worst unintended consequences." See Sheila Jasanoff, "Technology as a Site and Object of Politics," in *The Oxford Handbook of Contextual Political Analysis*, ed. Robert E. Goodin and Charles Tilly (Oxford: Oxford University Press, 2006), 746–47.

18. Von Kármán, *Toward New Horizons,* 5, 43, 60.

19. Michael B. Donley, Secretary of the Air Force, and General Norton A. Schwartz, US Air Force Chief of Staff, "Technology Horizons Study," Memorandum for ALMAJCOM-FOA-DRU/CC, 18 June 2009.

20. *Technology Horizons: A Vision for Air Force Science and Technology, 2010–2030*, Office of the US Air Force Chief Scientist, 15 May 2010, p. 123, http://www.defenseinnovationmarketplace.mil/resources/AF_TechnologyHorizons2010-2030.pdf.

21. Kolina Koltai et al., "Influence of Cultural, Organizational, and Automation Capability on Human Automation Trust: A Case Study of Auto-GCAS Experimental Test Pilots," International Conference on Human–Computer Interaction in Aerospace, Silicon Valley, CA, 30 July 2014.

22. *Technology Horizons*, 2010, xxi.

23. Scott Galster and Erica Johnson, "Sense-Assess-Augment: A Taxonomy for Human Effectiveness," Interim Report, Air Force Research Laboratory, 711 Human Performance Wing, Human Effectiveness Directorate, Wright-Patterson Air Force Base, OH, May 2013, p. 3.

24. Major Henry H. Arnold, "The Performance of Future Airplanes." 6 May 1925. RG 18, entry 22, box 502, National Archives, College Park, MD.

25. *Technology Horizons*, 2010, 73, 132.

26. Ibid., 73, 132, 156.

27. Ibid., 73.

28. Ibid.

29. *Autonomous Horizons: System Autonomy in the Air Force—A Path to the Future,* US Air Force Office of the Chief Scientist, June 2015.

30. *Autonomous Horizons* references the Air Force Research Laboratory's definitions of automation and autonomy. For automation, "The system functions with no/little human operator involvement; however, the system performance is limited to the specific actions it has been designed to do. Typically these are well-defined tasks that have predetermined responses (i.e., simple rule-based responses)." Autonomy describes a system that has "a set of intelligence-based capabilities that allow it to respond to situations that were not pre-programmed or anticipated in the design (i.e., decision-based responses). Autonomous systems have a degree of self-government and self-directed behavior (with the human's proxy for decisions)." As reflected in the substance of *Autonomous Horizons*, the AFRL further distinguishes between automation and autonomy: "In a static environment with a static mission, automation and autonomy converge. However, when dynamic missions take place in dynamic environments, automation can only support a small fraction of autonomy requirements. While the distinction is important, both have utility within various systems." See "Autonomy Science and Technology Strategy," Air Force Research Laboratory, Wright-Patterson Air Force Base, OH, 2013, http://www.defenseinnovationmarketplace.mil/resources/AFRL_AutonomyStrategy-DistroA.pdf.

31. *Autonomous Horizons*, 5.

32. Ibid., 6.

33. Ibid., 4.

34. Ibid., 5.

35. Author's discussion with Lockheed Martin F-35 cockpit simulator team, 14 April 2016, US Naval War College, Newport, RI.

36. *Autonomous Horizons*, 18.

37. Ibid., 10–11.

38. See Thomas B. Sheridan and William L. Verplank, "Human and Computer Control of Undersea Teleoperators," Man–Machine Systems Laboratory, Department of Mechanical Engineering, MIT, 14 July 1978. A simpler characterization of the spectrum of autonomy is portrayed in the US Department of Defense 2011 *Unmanned Systems Integrated Roadmap, FY2011-2036*. It identifies four levels where level 1, "Human Operated," applies when "a human operator makes all decisions. The system has no autonomous control of its environment although it may have information-only responses to sensed data." This contrasts with level 4, "Fully Autonomous," where "the system receives its goals from humans and translates them into tasks to be performed without human interaction. A human could still enter the loop in an emergency or change the goals, although in practice there may be significant time delays before human intervention occurs" (p. 46), http://www.fas.org/irp/program/collect/usroadmap2011.pdf. This concept is also discussed in greater detail in chapter 6's section on the automatic ground collision avoidance system. See also Caitlin Lee, "Embracing Autonomy: The Key to Developing a New Generation of Remotely Piloted Aircraft for Operations in Contested Air Environments," *Air and Space Power Journal* 25, no. 4 (Winter 2011).

39. *Autonomous Horizons*, 13. The report invokes the term *robustness* to describe "the degree to which the autonomy can sense, understand, and appropriately handle a wide range of conditions." It is the opposite of *brittleness*. For an early assessment of the capabilities of human and machine components, see Paul M. Fitts, ed., "Human Engineering for an Effective Air-Navigation and Traffic-Control System," a report to the Air Navigation Board and the National Research Council Committee on Aviation Psychology, March 1951.

40. Molly Lachance, Air Force Office of Scientific Research, "AFOSR Study Evaluates Trust in Autonomous Systems," 7 May 2015, http://www.wpafb.af.mil/news/story.asp?id=123447432.

41. F-35 pilot interview with author, 12 April 2016. See also Barrie Barber, "New Tech Tool at Wright-Patt Saving Lives," *Dayton Daily News*, 18 December 2015, http://www.mydayton dailynews.com/news/news/local-military/new-tech-tool-developed-at-wright-patt-saving-live /npnH4/.

42. Aaron M. Church, "The Science of Avoidance," *Air Force Magazine*, February 2016, 38.

43. *Autonomous Horizons*, 14, 19, 20.

44. Ibid., 14, 17, 20.

45. Ibid., 18.

46. Ibid., 11, 12, 15. Similarly, historian David Mindell uses the phrase "in the loop" to evaluate how NASA engineers integrated human and computer functions for the Apollo moon landing to accommodate astronaut desires for a greater amount of agency and control in a critical phase. David A. Mindell, *Digital Apollo: Human and Machine in Spaceflight* (Cambridge, MA: MIT Press, 2008), 14. For more on these terms regarding the human–machine control relationship, see Paul Scharre and Michael Horowitz, "An Introduction to Autonomy in Weapon Systems," Center for a New American Security, February 2015, 6.

47. *Autonomous Horizons*, 20.

48. The term *human–machine complementarity* is similarly used elsewhere to describe how "humans and machines help each other to achieve an effect of which each is separately incapable." H. H. Rosenbrock, ed. *Designing Human-Centered Technology: A Cross-Disciplinary Project in Computer-Aided Manufacturing* (London: Springer-Verlag, 1989), 38.

49. Sam LaGrone, "Mabus: F-35 Will Be 'Last Manned Strike Fighter' the Navy, Marines 'Will Ever Buy or Fly.'" *USNI News*, 15 April 2015, http://www.news.usni.org/2015/04/15/mabus -f-35c-will-be-last-manned-strike-fighter-the-navy-marines-will-ever-buy-or-fly.

50. Secretary of the Navy Ray Mabus speech to the National Press Club, 30 April 2015, http://www.c-span.org/video/?325629-1/us-navy-secretary-ray-mabus-remarks-national-press -club&start=1072.

51. Brendan Stickles (Commander, US Navy), "Twilight of Manned Flight?" *USNI Proceedings*, 142, no. 4 (April 2016): 28.

52. Melodie Feather, "The Northrup Grumman X-47B Unmanned Combat Air System Demonstration (UCAS-D) to Receive 2013 Robert J. Collier Trophy," National Aeronautic Association Press Release, 10 April 2014, http://www.naa.aero/userfiles/files/documents /Press%20Releases/Collier%20Recipient%202013%20PR.pdf.

53. National Aeronautic Association Collier Trophy Recipients, http://www.naa.aero/awards /awards-and-trophies/collier-trophy/collier-1911-1919-winners.

54. Megan Eckstein, "CNO: Navy Should Quickly Field CBARS to Ease Tanking Burden on Super Hornets," *USNI News*, 12 February 2016, http://www.news.usni.org/2016/02/12/cno -navy-should-quickly-field-cbars-to-ease-tanking-burden-on-super-hornets.

55. "A Navy in Balance? A Conversation with Chief of Naval Operations Admiral John Richardson," American Enterprise Institute interview, 12 February 2016, http://www.aei.org /events/a-navy-in-balance-a-conversation-with-chief-of-naval-operations-admiral-john -richardson/.

56. National Defense Authorization Act for FY 2016: Sec. 4201, "Research, Development, Test, and Evaluation," line 113, http://www.congress.gov/114/plaws/publ92/PLAW-114publ92 .pdf; Megan Eckstein, "House, Senate Armed Services Committees Agree to Support UCLASS, Additional Aircraft Procurement," *USNI News*, 29 September 2015, http://www.news.usni.org /2015/09/29/house-senate-armed-services-committees-agree-to-support-uclass-additional -aircraft-procurement.

57. The National Defense Authorization Act for fiscal year 2016 authorized drone pilots more flight pay ($1,000 vs. $850 monthly) and higher annual bonuses ($35,000 vs. $25,000)

than manned aircraft pilots ("officers performing other qualifying flight duty"). NDAA, Title VI, "Compensation and Other Personnel Benefits," Subtitle B, "Bonuses and Special and Incentives Pay," Section 617, p. 115. See also Brendan Stickles (Commander, US Navy), "Twilight of Manned Flight?" *Proceedings*, US Naval Institute 142, no. 4 (April 2016): 25.

58. "A Navy in Balance?"

59. US Air Force Global Hawk Fact Sheet, http://www.af.mil/AboutUs/FactSheets/Display/tabid/224/Article/104516/rq-4-global-hawk.aspx.

60. Geoffrey Summer et al., "The Global Hawk Unmanned Aerial Vehicle Acquisition Process," RAND Corporation report for the Defense Advanced Research Project Agency, 1997, http://www.rand.org/content/dam/rand/pubs/monograph_reports/2007/MR809.pdf.

61. Timothy Schultz, "UAS Manpower: Exploiting a New Paradigm," Air Force Research Institute, October 2009, p. 5.

62. "Secretary of Defense Gates' Speech at Air War College," 21 April 2008, http://www.cfr.org/world/secretary-defense-gates-speech-air-war-college/p16085.

63. Ibid.

64. Michael B. Donley, Secretary of the Air Force, address to the Air Force Association Air and Space Conference, Washington, DC, 15 September 2008. See also Schultz, "UAS Manpower," 2.

65. "United States Air Force RPA Vector: Vision and Enabling Concepts 2013–2038," February 2014, pp. 17–18, http://www.defenseinnovationmarketplace.mil/resources/USAF-RPA_VectorVisionEnablingConcepts2013-2038_ForPublicRelease.pdf.

66. US Air Force fact sheets, http://www.af.mil/AboutUs/FactSheets.aspx.

67. "United States Air Force RPA Vector," 18.

68. Schultz, "UAS Manpower," 5.

69. Peter Paret, Harmon Lecture Series: The Eighth Harmon Memorial Lecture in Military History, "Innovation and Reform in Warfare," USAFA, 1966, http://www.usafa.af.mil/df/dfh/harmonmemorial.cfm.

70. Schultz, "UAS Manpower," 8.

71. Ibid., 8, 22.

72. Houston R. Cantwell, "Beyond Butterflies: Predator and the Evolution of Unmanned Aerial Vehicles in Air Force Culture" (master's thesis, School of Advanced Air and Space Studies, 2007). See also Schultz, "UAS Manpower," 7.

73. Schultz, "UAS Manpower," 6.

74. Ibid. For more on this phenomenon and the history of the Predator and Reaper drones in general, see David Mindell, *Our Robots, Ourselves* (New York: Viking Press, 2015), 113–58, and Timothy Cullen, "The MQ-9 Reaper Remotely Piloted Aircraft: Humans and Machines in Action" (Ph.D. diss., Massachusetts Institute of Technology, 2011).

75. *Autonomous Horizons*, 18.

76. Schultz, "UAS Manpower," 8.

77. Ibid. Predator and Reaper crews may utilize "MIRC chat," or multiuser Internet relay chat, to facilitate brief, real-time, classified communication between the Ground Control Station and various end-users in the combat area of operations, to include troops in contact with the enemy as well as entities such as the Air Operations Center. Similar to manned aircraft using line-of-sight radios, RPA crews utilize satellite communication to speak directly with ground troops who are receiving drone-fed information and, in turn, identifying potential targets for drone strikes. See also Mindell, *Our Robots, Ourselves*, 145–46.

78. Schultz, "UAS Manpower," 8.

79. According to Colonel Stephen Wilson, Air Education and Training Command assistant operations officer, there are three tiers of categorization. Tier I identifies top-notch performers

the commander seeks to retain; tier II represents otherwise excellent performers who don't quite make tier I; tier III identifies competent but average performers. Commanders are much more inclined to send someone from tier III than tier I or II to a nonvolunteer assignment. Schultz, "UAS Manpower," 20, and Houston R. Cantwell, Lieutenant Colonel, USAF, "Air Force Unmanned System Operators: Breaking Paradigms," *Air and Space Power Journal* 23, no. 2 (January 2009): 67–77. See also Government Accountability Office (GAO) Report to Congressional Requesters, GAO-14-316, "Air Force RPA Pilots," April 2014, p. 33.

80. GAO Report, "Air Force RPA Pilots," 18. For more on attitudes within the RPA community, see Michael W. Byrnes (Captain, USAF), "Dark Horizon: Airpower Revolution on a Razor's Edge—Part Two of the 'Nightfall' Series," *Air and Space Power Journal* 29, no. 5 (September–October 2015): 31–56, and Mindell, *Our Robots, Ourselves*, 152–57.

81. Mindell, *Our Robots, Ourselves*, 152–54.

82. GAO Report, "Air Force RPA Pilots," 12, 13.

83. Ibid. See also Herbert J. Carlisle (General, USAF), "Remotely Piloted Aircraft Enterprise," statement to the Senate Armed Services Committee Subcommittee on Airland, 16 March 2016, p. 5, http://www.armed-services.senate.gov/imo/media/doc/Carlisle_03-16-16.pdf.

84. Air Force Instruction 36-101, "Classifying Military Personnel (Officer and Enlisted)," 25 June 2013, 70.

85. GAO Report, "Air Force RPA Pilots," 7.

86. Initial Flight Training (IFT) Pre-arrival Guide, Version 23, April 2016, Doss Aviation, Pueblo, CO, http://www.dossifs.com/docs/task1/Arrival_Guide.pdf.

87. James A. Whitmore, Major General, USAF, "Moving Forward: The Next Generation of Combat Aviators," briefing from the Air Education and Training Command, http://www.dtic.mil/ndia/2010targets/Whitmore.pdf.

88. GAO Report, "Air Force RPA Pilots," 18.

89. Carlisle, "Remotely Piloted Aircraft Enterprise," 5. See also David DeKunder, "Air Force RPA Training Pipeline Set to Expand," Public Affairs Office, Joint Base San Antonio, 20 August 2015, http://www.jbsa.mil/DesktopModules/ArticleCS/Print.aspx?PortalId=102&ModuleId=25294&Article=61397.

90. "AF Announces FY16 Aviator Bonuses: RPA Pilots Eligible for First Time," Secretary of the Air Force Public Affairs Command Information, 15 December 2015. Officers who completed undergraduate RPA training (18X pilots) plus six years of aviation service are eligible for the bonus and may opt for half up front. http://www.af.mil/DesktopModules/ArticleCS/Print.aspx?PortalId=1&ModuleId=850&Article=636358.

91. GAO Report, "Air Force RPA Pilots," 52.

92. Churchill Eisenhart, "Cybernetics: A New Discipline," *Science* 109, no. 2834 (22 April 1949): 397; Schultz, "UAS Manpower," 28.

CONCLUSION: The Past and Future of Pilots

1. Kevin M. Henry, personal correspondence, 24 October 2006.

2. Kevin M. Henry, telephone interview by author, 11 April 2007. Although he permanently lost his flying status, Kevin Henry's neurological injuries did not preclude a full, normal lifestyle. See also S. L. Jersey et al. "Severe Neurological Decompression Sickness in a U-2 Pilot," *Aviation, Space, and Environmental Medicine* 81, no.1 (January 2010): 64–68.

3. Detlev W. Bronk et al., eds., *Advances in Military Medicine Made by American Investigators Working under the Sponsorship of the Committee on Medical Research* (Boston: Little, Brown, 1948), 207.

4. Harry G. Armstrong, "The Influence of Aviation Medicine on Aircraft Design and Operation," *Journal of the Aeronautical Sciences* 5, no. 5 (March 1938): 193–98; L. H. Bauer,

"Aeronautics and the Practice of Medicine," *Journal of Aviation Medicine* 1 (March–December 1930): 81.

5. Armstrong, "The Influence of Aviation Medicine," 193.

6. "Basic Instrument Flying," *U.S. Army Air Forces Technical Order*, no. 30-100A-1 (1 June 1943): 1–2. File G-2, box 851, USAF Museum, Colorado Springs, CO.

7. Norbert Wiener, *Cybernetics, or Control and Communication in the Animal and the Machine* (New York: John Wiley and Sons, 1948), 38.

8. David Mindell, *Digital Apollo: Human and Machine in Spaceflight* (Cambridge, MA: MIT Press, 2008), 26. The necessity for pilots to develop new skills due to advances in automation is also discussed by Raymond P. O'Mara, "The Socio-technical Construction of Precision Bombing: A Study of Shared Control and Cognition by Humans, Machines, and Doctrine during World War II" (Ph.D. diss., Massachusetts Institute of Technology, 2011).

9. Mindell, *Digital Apollo*, 24.

10. Ibid., 5, 28.

11. Norbert Wiener, the father of cybernetics, observed in 1942 that "the pilot behaves like a servo-mechanism" as he effected control of the aircraft by integrating inputs based on his own senses and the feedback received from the aircraft. Peter Galison, "The Ontology of the Enemy: Norbert Wiener and the Cybernetic Vision," *Critical Inquiry* 21, no. 1 (Autumn 1994): 236. Similarly, David Mindell uses the term *manual servomechanism* to describe how the gunner in an antiaircraft artillery crew was part of a machine-like apparatus. David Mindell, *Between Human and Machine: Feedback, Control, and Computing before Cybernetics* (Baltimore: Johns Hopkins University Press, 2002), 92–93, 98.

12. The Aero Medical Laboratory, *Your Body in Flight: An Illustrated "Book of Knowledge" for the Flyer* (Patterson Field, OH: Air Service Command, 20 July 1943), 7. Manuscript series 45, box 23, Clark Special Collections Branch, USAFA Library, Colorado Springs, CO (emphasis original).

13. Norton Schwartz, General, USAF, Air Force Association Convention Speech, 15 September 2009, Washington DC.

14. Lewis Mumford, "Authoritarian and Democratic Technics," *Technology and Culture* 5, no. 1 (Winter 1964): 5.

15. Sheila Jasanoff, "Technology as a Site and Object of Politics," in *The Oxford Handbook of Contextual Political Analysis*, ed. Robert E. Goodin and Charles Tilly (Oxford: Oxford University Press, 2006), 746.

16. Mumford, "Authoritarian and Democratic Technics," 5.

17. Sungook Hong, "Man and Machine in the 1960s," *Techne* 7, no. 3 (Spring 2004): 49–77.

18. Mumford, "Authoritarian and Democratic Technics," 2, 7. For more on the social construction of technology concept, see Thomas P. Hughes, *Networks of Power* (Baltimore: Johns Hopkins University Press, 1983) and *Human-Built World: How to Think about Technology and Culture* (Chicago: University of Chicago Press, 2005); also Wiebe E. Bijker, Thomas P. Hughes, and Trevor Pinch, eds., *The Social Construction of Technological Systems: New Directions in the Sociology and History of Technology* (Cambridge, MA: MIT Press, 2012).

19. This concept, typically referred to as the social construction of technological systems, is explored in Bijker et al., *The Social Construction of Technological Systems*.

20. In their assertion of influence and control, some contend that pilots, including several senior leaders in the Air Force, routinely acted to stall the institution's transition to a greater number of unmanned aircraft that may infringe on the pilot-centric status quo. In a manner similar to how incumbent commercial enterprises sometimes fail to adapt, these status quo proponents tend to "dismiss, disparage, and then bargain" regarding the efficacy of new

technology. The bargaining occurs when incumbents "insist that new weapons [i.e., unmanned aircraft] are best used in conjunction with existing ones." For this and other insights on adoption of innovations in the Air Force, see Lawrence Spinetta, "Remote Possibilities: Explaining Innovations in Airpower" (Ph.D. diss., School of Advanced Air and Space Studies, Air University, Maxwell Air Force Base, AL), 260–64.

21. Christian Davenport, "Meet the Most Fascinating Part of the F-35: The $400,000 Helmet," *Washington Post*, 1 April 2015.

22. H. R. Everett, *Unmanned Systems of World Wars I and II* (Cambridge, MA: MIT Press, 2015), 248–60.

23. Tom Ehrhard, *Air Force UAVs: The Secret History* (Arlington, VA: Mitchell Institute Press, 2010), 2, 4. Ehrhard offers an insightful history of unmanned systems developed and employed by the Air Force after the Second World War. For a more detailed and comprehensive appraisal, see Tom Ehrhard, "Unmanned Aerial Vehicles in the United States Armed Services: A Comparative Study of Weapon System Innovation" (Ph.D. diss., Johns Hopkins University, 2001).

24. Donald MacKenzie asserts, "It was not surprising [during the early Cold War] . . . that an organization dominated by pilots, as the Air Force was (and to a substantial extent still is), should be reluctant to see its central strategic role filled by anything other than a manned system." Donald MacKenzie, *Inventing Accuracy: A Historical Sociology of Nuclear Missile Guidance* (Cambridge, MA: MIT Press, 1993), 101.

25. Richard Clark (Lieutenant Colonel, USAF), "Uninhabited Combat Aerial Vehicles," *CADRE Paper No. 8* (Maxwell Air Force Base, AL: Air University Press, 2000), 28–34. Clark notes that one obstacle to unmanned systems development may have been a pro-pilot bias, yet he acknowledges that "the reluctance of the services to embrace UCAVs was not based on the threat to the status of pilots and manned aircraft but on the Air Force leadership's skepticism towards the effectiveness of UCAVs. This skepticism may indicate that within the Air Force culture there is an aversion to taking risks on unmanned aviation technology until the uncertainty is reduced significantly. Whether the reluctance towards unmanned systems is based on risk aversion or based on the maintenance of the status of pilots, it is an obstacle inherent in the Air Force's culture that must be addressed" (32). For additional perspective on unmanned aircraft and Clark's analysis, see James A. Sweeney, "The Wave of the Present: Remotely-Piloted Aircraft in Air Force Culture" (master's thesis, School of Advanced Air and Space Studies, Air University, Maxwell Air Force Base, AL).

26. David Mindell, *Our Robots, Ourselves* (New York: Viking Press, 2015), 125.

27. Theodore von Kármán, *Where We Stand: A Report of the AAF Scientific Advisory Group*, 1945, 48–49.

28. See http://www.northropgrumman.com/Capabilities/anapg81aesaradar/Pages/default.aspx.

29. Briefing to author by Lockheed Martin F-35 cockpit simulator team, 14 April 2016, US Naval War College, Newport, RI.

30. Ibid.

31. Dr. Greg L. Zacharias, "Advancing the Science and Acceptance of Autonomy for Future Defense Systems," presentation to the House Armed Services Committee Subcommittee on Emerging Threats and Capabilities (US. House of Representatives, 19 November 2015), 5, https://www.docs.house.gov/meetings/AS/AS26/20151119/104186/HHRG-114-AS26-Wstate-ZachariasG-20151119.pdf.

32. Zacharias, "Advancing the Science and Acceptance of Autonomy" (emphasis added).

33. Timothy Schultz, "UAS Manpower: Exploiting a New Paradigm," Air Force Research Institute, October 2009, 10.

34. Michael Hoffman, "USAF Chief: Crews Will Operate Multiple UAVs," *Defense News*, 22 (May 2009).

35. Nicholas Ernst et al., "Genetic Fuzzy Based Artificial Intelligence for Unmanned Combat Aerial Vehicle Control in Simulated Air Combat Missions," *Journal of Defense Management* 6, no. 1 (2016): 1–7.

36. Matthew E. Rowan et al., "Flight Test Evaluation of the AFRL Automated Wingman Functions (Project HAVE Raider)," Final Technical Information Memorandum 14B-01, June 2015, USAF Test Pilot School, Edwards Air Force Base, CA.

37. "U.S. Air Force, Lockheed Martin Demonstrate Manned/Unmanned Teaming," Lockheed Martin Press Release, 10 April 2017, http://news.lockheedmartin.com/2017-04-10-U-S-Air-Force-Lockheed-Martin-Demonstrate-Manned-Unmanned-Teaming.

38. Chesley Sullenberger and Jeffrey Zaslow, *Highest Duty: My Search for What Really Matters* (New York: HarperCollins, 2009), 190.

39. The National Advisory Committee for Aeronautics, Special Committee on the Nomenclature, Subdivision, and Classification of Aircraft Accidents, "The Classification of Aircraft Accidents: Nomenclature and Subdivision," 11 June 1928, 6–7. Call number 200.3912-2, iris number 00141796, Air Force Historical Research Agency (AFHRA).

40. "Operational Use of Flight Path Management Systems," Final Report of the Performance-based Operations Aviation Rulemaking Committee / Commercial Aviation Safety Team Flight Deck Automation Working Group, Federal Aviation Administration, 5 September 2013, pp. 2, 3, http://www.faa.gov/about/office_org/headquarters_offices/avs/offices/afs/afs400/parc/parc_reco/media/2013/130908_PARC_FltDAWG_Final_Report_Recommendations.pdf. For additional analysis, see Mindell, *Our Robots, Ourselves*, 73–77.

41. David D. Woods and Nadine B. Sarter, "Learning from Automation Surprises and 'Going Sour' Accidents: Progress on Human-Centered Automation." Final Report, NASA Ames Research Center, 19 January 1998, 6, 21, http://www.researchgate.net/publication/24324128_Learning_from_Automation_Surprises_and_Going_Sour_Accidents_Progress_on_Human-Centered_Automation.

42. "Final Report on the Accident on 1st June 2009 to the Airbus A330-203 Registered F-GZCP Operated by Air France, Flight AF 447 Rio de Janeiro—Paris," Bureau d'Enquêtes et d'Analyses pour la sécurité de l'aviation civile, 27 July 2012, p. 199.

43. Wiener, *Cybernetics*, 19.

44. Norbert Wiener, "Moral and Technical Consequences of Automation," *Science* 131, no. 3410 (6 May 1960): 1355.

45. Ibid., 1358.

46. Ibid.

47. This reflects historian Lynn White's notable claim that a new technical ability "merely opens a door; it does not compel one to enter. The acceptance or rejection of an invention, or the extent to which its implications are realized if it is accepted, depends quite as much upon the condition of a society, and upon the imagination of its leaders, as upon the nature of the technological item itself." Lynn White Jr., *Medieval Technology and Social Change* (London: Oxford University Press, 1968), 28.

Coda

1. For a comprehensive account of this accident, see National Transportation Safety Board Accident Narrative no. NYC99MA178, 7 June 2000.

2. Ibid., 15.

3. Ibid., 1.

4. John Kennedy Jr., completed twelve of twenty-five lessons required to obtain an instrument rating. Without an instrument rating, his pilot license permitted him to fly only in "visual meteorological conditions," but not in "instrument meteorological conditions" (or blind flight). One of his certified flight instructors stated that Kennedy had not completed enough training to pass the flying evaluation required for an instrument rating. The same CFI also told Kennedy he could fly with him to Martha's Vineyard that night, but Kennedy stated he "wanted to do it alone." Ibid., 4.

Index

ABOUT THE AUTHOR

This book results from Timothy P. Schultz's unique experience as an aviator and historian. During his twenty-six years in the US Air Force, Schultz served as an instructor pilot in a variety of aircraft, including the supersonic T-38 and the high-altitude / near space reconnaissance U-2. He combined his diverse aviation background and worldwide operational experience with master's degrees in cellular biology (Colorado State University), military art and science (Air Command and Staff College), and airpower art and science (School of Advanced Air and Space Studies). He also earned a PhD in history from Duke University, where he studied our evolving relationship with technology. Formerly the commandant and dean of the US Air Force's School of Advanced Air and Space Studies, Schultz now serves as the associate dean of Academics at the US Naval War College.